A WORLD OF WOUNDS

A WORLD OF WOUNDS

Rebuilding a
Bipartisan Environmental Movement
and
Cultivating Authentic Hope

Nancy J. Manring

STANFORD UNIVERSITY PRESS
Stanford, California

Stanford University Press
Stanford, California

Library of Congress Cataloging-in-Publication Data
Names: Manring, Nancy author
Title: A world of wounds : rebuilding a bipartisan environmental movement and cultivating authentic hope / Nancy J. Manring.
Description: Stanford, California : Stanford University Press, [2025] | Includes bibliographical references and index.
Identifiers: LCCN 2025008446 (print) | LCCN 2025008447 (ebook) | ISBN 9781503643659 cloth | ISBN 9781503644403 paperback | ISBN 9781503644410 ebook
Subjects: LCSH: Environmentalism—United States—History | Environmental policy—United States
Classification: LCC GE197 .M27 2025 (print) | LCC GE197 (ebook) | DDC 363.700973—dc23/eng/20250428

LC record available at https://lccn.loc.gov/2025008446
LC ebook record available at https://lccn.loc.gov/2025008447

Cover design: Daniel Benneworth-Gray
Cover art: Unsplash and Flickr

The authorized representative in the EU for product safety and compliance is: Mare Nostrum Group B.V. | Mauritskade 21D | 1091 GC Amsterdam | The Netherlands | Email address: gpsr@mare-nostrum.co.uk | KVK chamber of commerce number: 96249943

For Evan and Emma

Contents

Preface

Imagine wondering if the physical world is real. Growing up in a family fascinated by spirituality, metaphysics, and popular conceptions of theoretical physics, I was taught to question the nature of reality. Some in my family were drawn to Einstein's writings (my maternal grandfather bore a striking resemblance to him, minus the unruly hair). Einstein once declared, "Concerning matter, we have all been wrong. . . . There is no matter."[1] The fact that atoms consist of mostly empty space was offered as evidence that reality is not actually material. As a child, I grappled with this puzzling notion. I can still recall standing in our backyard, scrutinizing a leaf—a ten-year-old trying to make sense of her world. Was the leathery, almost weightless leaf in my hand real? As I wandered around my mother's backyard garden, nibbling green apples, I tried to wrap my head around the possibility that the earth—all that I could see and touch and taste—was not actually material. I'd imagine myself perched high in my favorite apple tree only to have it collapse under me as it lost all substance. How could this be true?

And yet, growing up in a family that also loved nature, I was constantly immersed in the reality of the natural world. Family conversations often began with nature bulletins recounted regularly like weather reports, rich with anecdotes about backyard birds and beautiful sunsets. All year long, we hiked and canoed in a nearby state park, often visiting the cool grotto filled with ferns and moss, home of a secret natural spring. Sweeping aside floating leaves, we could drink clear, cold water right out of the earth. In winter, we skated on winding streams, hiked to our favorite waterfall to see the living water transformed into a blue-white ice stalagmite, and got faces full of snow as we raced downhill on our toboggan. Twice our summer vacations took us out west to some of our most spectacular national parks. Memories of red rock hoodoos in Bryce, beckoning box canyons in Zion, the enormity and subtle palette of colors of the Grand Canyon, tumbling waterfalls in Yosemite, and the pungent, sulfurous smells and aquamarine geysers of Yellowstone became part of me.

Not surprisingly, in college I was drawn to the natural sciences, studying horticulture, plant biology, and ecological systems theory. Eventually my childhood story, education, and research converged. During my doctoral research in natural resource policy, I was introduced to the centrality of clashing worldviews as drivers of conflict in environmental disputes. Like putting on a special pair of glasses, worldviews provide a lens for perceiving and interpreting "how the world works." Worldviews (also known as social paradigms) contain implicit assumptions about the nature of the physical world and the severity of our environmental problems. Having been raised in a family fascinated by alternative worldviews, the notion of conflicting worldviews struck a resonant, deeply familiar chord for me. I quickly grasped how clashing worldviews—with profoundly contrasting views of nature—shape environmental politics. Armed with an ecological education, I realized society's dominant worldview offers a mythological interpretation of the natural world. Just as I was taught to question the physical reality of the material world, the dominant worldview in American society has led large numbers of people to question the reality of contemporary environmental degradation. Seduced by the dominant worldview's assurances of human exemptionalism and superiority, vast numbers of Americans believe humans are immune to climate change and ecological degradation.

After receiving my doctorate from a School of Natural Resources, I joined a political science department.[2] Although I work primarily with social scientists, my ontological assumptions are shaped by the natural sciences and grounded in the physical materiality of the earth. I soon encountered the fact that natural scientists' ontological claims often collide with those of social scientists. As Johannes Persson et al. explain,

> Social and natural scientists tend to approach questions of truth and objective reality differently. Many social scientists emphasize that all

human representations of reality are contingent constructions, and some would even argue that it is meaningless to speak of an objective reality. Natural scientists, on the other hand, often visualize their research results and conclusions as accurate, or approximately accurate, representations of a posited reality.[3]

When considering nature, social scientists known as social constructionists typically focus on the social construction of nature, in other words, *perceptions* of nature. Some constructionists argue that the natural world itself "is a social construction that does not exist independently of human thinking."[4] Having grown up immersed in nature, I found this assertion striking. I have hiked for miles on earthen trails, hugged massive trees, and camped on hard, damp ground. I have experienced the somnolent parching heat of the desert, the numbing waters of mountain streams, the taste of briny ocean water, and the dizzying sensation of receding waves carrying the sand under my toes back to the sea. And to this day, I am entranced by the astonishing beauty of nature, from the tiniest mosses and wildflowers to the towering peaks of the Rockies. Human thinking did not conjure the wild and wonderful reality of the natural world. The natural world exists independently of human interpretations and perceptions. Unfortunately, the natural world is not immune to human impacts.

As a scholar trained in both the social and natural sciences, I focus on the physical realities of nature and the impacts of human social systems on nature. I believe the massive body of evidence produced by thousands of natural scientists in multiple disciplines is an accurate representation of the physical world and our environmental problems. In other words, the ontological perspective that shapes this book is influenced by the reality, urgency, and growing danger of contemporary planetary challenges. The picture of our planet revealed by thousands of natural scientists' decades of research is a world in danger of catastrophic climate change, unraveling ecological systems, and planetary destabilization.

However, the planetary story revealed by contemporary scientists is not a new story. In 1992, well over three decades ago, 1,700 of the world's top scientists, including most of the Nobel laureates in the sciences at the time, issued a letter to humanity with a stark warning:

Human beings and the natural world are on a collision course. Human activities inflict harsh and often irreversible damage on the environment and on critical resources. If not checked, many of our current practices put at serious risk the future that we wish for human society ... and may so alter the living world that it will be unable to sustain life in the manner that we know. Fundamental changes are urgent if we are to avoid the collision our present course will bring about.[5]

We did not listen.

Twenty-five years later, 15,364 scientists from 184 countries issued another dire warning: "Humanity has failed to make sufficient progress in generally solving these foreseen environmental challenges, and alarmingly, most of them are getting worse. Especially troubling is the current trajectory of potentially catastrophic climate change."[6] In 2021, seventeen notable scientists published an article titled "Underestimating the Challenges of Avoiding a Ghastly Future," in which they argued, "The scale of the threats to the biosphere and all its lifeforms—including humanity—is in fact so great that it is difficult to grasp for even well-informed experts."[7]

Still, most of us did not listen.

In 2022, the Intergovernmental Panel on Climate Change (IPCC) stressed, "The evidence is clear: the time for action is now. . . . The decisions we make now can secure a livable future."[8] Although American politics typically are frustratingly slow and incremental, we do not have the luxury of delay in environmental politics. We need to move forward based on the massive body of scientific evidence produced by Earth System scientists and other natural scientists in a host of disciplines. We need to respond to the clarion call for action issued by literally thousands of scientists all across the world. What does it say when dedicated climate scientists, so exasperated by political inaction, call on their colleagues to "stage a mass walkout, to stop their research until nations take action on global warming?"[9] We need to take scientists' warnings very seriously.

Heeding the scientists' warnings means that we need to rethink the role of science in environmental politics and policymaking. Years ago, Kai Lee argued that we need the compass of science to guide democratic deliberations in environmental policymaking.[10] However, environmental science must be more than a tool; it must be the foundation of environmental debate and policymaking. Following William Lafferty and Eivind Hovden, I maintain that the natural sciences should be given "principled priority" in environmental politics and policy.[11] The environmental issues that constitute environmental politics are unlike other political issues. Environmental issues comprise more than values, ideology, and opinions:

environmental issues are grounded in the physical sciences and the earth. Social science research informs policymaking in many areas of American politics; however, we don't give "principled priority" to social scientists' findings. We prioritize values, ideology, and economics. Moreover, when it comes to controversial political issues such as abortion or immigration or taxes, there are always two sides. In fact, like the one hundred-armed Hecatoncheire giants of ancient Greek mythology, in contemporary politics there are countless sides and positions. However, when it comes to the state of the earth, there is just one side. Environmental science is not just one perspective in the messy world of environmental politics.

In advocating for the central role of science in environmental politics and policy, it is important to distinguish between the role of science in our politics and the use of science in environmental policymaking. In contemporary environmental politics, we have an informational free-for-all. All kinds of specious claims and scientific distortions about climate change, biodiversity loss, and other major environmental problems zoom through cyberspace and fill the airwaves, confusing the public and preventing political mobilization for change. How can we expect people to be informed, responsible citizens in a world awash in scientific misinformation? We have tolerated the distortion of science in environmental politics for too long.

The role of science in environmental policymaking is equally important but more complicated. We cannot simply supplant democratic dialogue with science-based decision-making in environmental policy deliberations. My views on environmental policymaking have been shaped by my doctoral training in environmental conflict and dispute resolution. Disputes over data are legendary in environmental disputes and policymaking. Dispute resolution scholar and practitioner Lawrence Susskind refers to the "battle of the printout" in environmental disputes.[12] Similarly, political scientist Walter Rosenbaum notes that "data become weapons"[13] in environmental politics. Environmental mediators have introduced innovative processes collectively referred to as data negotiation for dealing with disputes over environmental science in site-specific environmental disputes and policy dialogues. In the data negotiation process, stakeholders agree on the sources of scientific information used to inform their deliberations. Disputants occasionally design needed studies together, agreeing on the geographic scope, time frame, and key variables to be examined. Just as credible science must be the foundation of environmental dispute resolution efforts, credible science also must be foundational in environmental policymaking. Fortunately, there is a growing literature based in research and practice that demonstrates it is possible to balance science and democracy in policymaking processes through collaborative deliberation informed by accurate scientific information. Our political and environmental well-being—and perhaps even our ability to thrive as a democratic nation in a world of environmental losses—depend on better integrating science and democratic norms.

My views on the role of science in environmental politics also are informed by the fact that environmental policy decisions often pose unique challenges unlike other areas of social policy. Policy reversibility is the norm in democratic politics; unpopular decisions can be reversed in the next presidential administration or Congress. However, environmental policy decisions can lead to irrevocable ecological losses that cannot be reversed on any meaningful human time scale. Irreversible losses contravene democratic norms of policy reversibility and rob living and future generations of their ecological inheritance. Every day as I look into the faces of my students, a montage of irreversible environmental losses and ever-worsening threats that will compromise the quality and prospects of their adult lives runs through my mind. For the sake of our children, we must listen to the scientists.

However, the story of environmental politics offered by this book is a story not only grounded in scientific realities, it is also grounded in *political* realities. In most conventional analyses of environmental politics, longstanding norms of bipartisanship lead to false equivalencies that obscure important political dynamics, causing widespread frustration and even despair. Perhaps more importantly, the conventional wisdom about partisanship in American society ignores the fact that the partisan rift in environmental politics has been purposefully and artificially cultivated by conservative opponents of the environmental movement. This contrived partisan divide obscures underlying interests that unite people of all political persuasions. It is my hope that debunking false assertions about partisanship in environmental politics will help readers come back home to the reality of Americans' shared love of our natural heritage. Fortified by a fresh story of environmental politics, authentic hope, and effective communication tools, it is my hope that readers will reach across the manufactured partisan divide and join their families, neighbors, and communities in rebuilding a bipartisan environmental movement working to protect our planetary home. There is no more urgent task before us.

Acknowledgments

This book represents an intellectual journey of decades. Numerous teachers, advisors, colleagues, friends, and family members have shaped my journey. At Ohio State University, the late Dr. Gareth Gilbert taught me to "think in systems," a way of seeing the world that has informed my work from my master's degree to the present. My graduate advisor at the University of Michigan, Dr. James Crowfoot, introduced me to the power of collaborative dispute resolution and the research on clashing worldviews, perspectives that have guided my teaching of environmental politics and informed the writing of this book.

Thanks go to my inspiring colleagues and mentors in the Department of Political Science at Ohio University, especially Lysa Burnier, Barry Tadlock, Nukhet Sandal, Susan Burgess, and the late Patricia Weitsman. I also have been fortunate to work closely with colleagues from other departments at Ohio University who have enriched my perspectives on environmental sustainability including Loraine McCosker, Edna Wangui, Kelly Johnson, Viorel Popescu, Harold Perkins, Ryan Fogt, Amy Lynch, Dina Lopez, and Alyssa Bernstein.

To the women in the Resist/Persist Book Club, I am grateful for their friendship, inspiration, and interest in my work as well as for their unwavering advocacy of democracy and social justice. My good friend Austin Babrow helped me strengthen my argument, offered advice on academic publishing, and gave me a boost when my confidence was flagging. Thanks also go to my sister support team, Susan Manring and Joan Hull; I'm so grateful for their support, interest in my work, and tireless editing help.

To my husband Eric Fenstermaker, my companion and "supporter in chief," I owe my deepest thanks. He understood that writing a book is a long process that requires sacrifices from all. He graciously gave up time together when my ideas were flowing and editing was all-consuming. His editorial comments strengthened my writing, and his love and laughter provided respite from the rigors of academic life.

Special thanks go to my editor, Dan LoPreto, for his advocacy of my work, and the guidance he provided as he shepherded my book through the review processes. Thanks also go to the editorial team including Thane Hale, Emily Smith, and Jennifer Gordon, all of whom helped to convert my manuscript into this book. Lastly, deepest gratitude goes to my late parents, Jane and Lawson Manring, who instilled in me a fierce love of nature and passion for lifelong learning.

A WORLD OF WOUNDS

INTRODUCTION

Americans need a new story of environmental politics. We need a hopeful story that matches the demands of two interwoven existential threats: the realities of contemporary global environmental problems and threats to American democracy. Like the rest of the global community, we are facing accelerating global warming, biodiversity loss, and novel forms of pollution that endanger human welfare and planetary stability. Americans are frustrated and worried about the continued viability of a political system rocked by extreme polarization. In a nation rife with partisanship, where is a new story of environmental politics to be found? A new story of environmental politics has its roots in our history. We have forgotten the power of the environment to bring people together.

In 1962, Rachel Carson's landmark book *Silent Spring* captured the American imagination.[1] *Silent Spring* told the story of how DDT and other chemically similar pesticides persist in the environment, causing widespread ecological damage. Beginning in the mid-1940s, DDT was used extensively in agriculture and in communities across the United States to control mosquitos and household pests. Clouds of DDT were sprayed over the heads of children and families at the beach, in swimming pools, and picnicking in public parks. However, DDT also killed songbirds, butterflies, bees, and fish. As birds of prey at the top of the food chain consumed contaminated fish, DDT became concentrated in their fatty tissues causing their eggshells to be so thin that they immediately broke when laid, scrambled eggs instead of a healthy clutch. Already threatened by habitat loss, DDT caused a precipitous drop in the Bald Eagle population; by 1963, our national symbol was teetering on the edge of extinction.

Not long after the release of *Silent Spring,* a string of well-publicized disasters in the late 1960s sent more shockwaves through the American public. Despoiled by unchecked pollution and widespread algae blooms, Lake Erie—the twelfth-largest freshwater lake in the world—was pronounced "dead." The pristine Santa Barbara coast was tarred by a massive oil spill, and Cleveland's Cuyahoga River caught on fire. Countless rivers and streams were polluted and choked with litter, and Americans could see, smell, and sometimes even taste air pollution. According to one observer, "Alarm about the environment sprang from nowhere to major proportions in a few short years."[2]

Americans' burgeoning environmental worries coincided with the tumultuous politics of the civil rights movement and the Vietnam War. In the midst of political unrest and social division, concern for the environment became a unifier. Convinced that Americans' environmental concerns far outpaced current political leadership, Senator Gaylord Nelson conceived the idea of Earth Day. Inspired by anti-war protests, he wanted an environmental demonstration so big that Congress couldn't ignore it. On April 22, 1970, an estimated 20 million people all across the political spectrum participated in the first Earth Day, rallying for environmental protection. The extraordinary growth of public support for environmentalism paired with bipartisan political leadership led to the creation of the U.S. Environmental Protection Agency and the passage of an unprecedented number of landmark environmental statutes in the early 1970s including the Clean Air Act,[3] Clean Water Act,[4] Safe Drinking Water Act,[5] and Endangered Species Act.[6] As the 1970s and 1980s unfolded, Americans were introduced to a string of new hazards ranging from toxic wastes to acid rain, the ozone hole, and global warming. By 1989, three-fourths of Americans identified themselves as environmentalists.

Fast-forward to the present. Scientists have announced that global warming is accelerating, and the impacts of climate change have become impossible to ignore as a growing roster of communities are devastated by catastrophic floods, fires, and record heatwaves. Biologists warn of a sixth mass extinction unprecedented in speed, unraveling the ecological fabric of life. Toxic "forever chemicals" are seemingly everywhere, and scientists report that it is raining plastic. Yet by 2021, the percentage of Americans who self-identify as environmentalists has fallen sharply. Today only 41 percent of Americans called themselves environmentalists. What happened to the popular environmental movement that catalyzed social and political change decades ago?

Pollsters report that a widening partisan gulf has emerged between conservatives and liberals, particularly around environmental issues and climate change.[7] Media coverage reinforces the perceived durability of entrenched positions. Afraid of controversy, people rarely discuss climate change and environmental issues with family and friends. How did concern for the environment become a partisan issue? Is the declining popularity of the environmental movement simply a casualty of time or newer political concerns? The partisan divide in environmental politics did not emerge organically in American politics: the partisan rift was purposely constructed by American conservatives to undermine the power of the environmental movement. This artificial partisan gulf has alienated Americans from one another and masked Americans' environmental concerns and traditional love of our natural heritage.

Weary of partisan politics, countless Americans are yearning for political unity, bipartisan solutions, and hope. According to the Pew Research Center, 79 percent of Americans are frustrated "there is so much political disagreement" over climate change.[8] The current story of environmental politics—with its false equivalencies and bipartisan failures—has failed the American people. Typical accounts offer a story of conflicting political ideologies and political stalemate that has blocked the passage of essential environmental legislation. The usual story is one of partisan division that has alienated families, neighbors, and friends. It is a story of futility that discourages many people from adopting environmentally friendly behaviors. It is a story of frustration and hopelessness that drives people away from politics at a time when the need for political engagement could not be more pressing.

Individuals may choose to ignore politics, but we cannot ignore planetary realities. Protecting our common planetary home requires all of us, Democrats and Republicans alike. It is time to bridge the partisan divide designed to separate the American people from each other and the environment. It is time for a fresh, bipartisan story of environmental politics. We need a hopeful story of environmental politics that inspires and empowers people to vote, work and advocate for the protection of America's natural heritage and our planetary home. Reconnecting with Americans' longstanding love of nature and concern for environmental quality provides the foundation for a hopeful story of environmental politics. Underneath what appear to be entrenched politically charged positions are a variety of widely shared interests and environmental concerns. Americans across the political spectrum care about clean air and water, wildlife conservation, public lands and outdoor recreation, all of which are adversely affected by fossil fuel extraction and global warming. It is time to rebuild the story of environmental politics in the United States.

ENVIRONMENTAL POLITICS AND EMOTION

A story of environmental politics that motivates people to act in defense of our planetary home must be based on authentic hope. We must tell the honest story of our planetary emergency and expose the actual story of environmental politics in the United States. But what do we mean by "authentic" hope? Hope has many connotations and often is likened to optimism. Optimism and authentic hope are not the same; optimism can be lazy and deluded. As noted environmental studies scholar David Orr emphasizes,

> Optimism leans back, puts its feet up, and wears a confident look knowing that the deck is stacked. . . . Hope requires us to check our optimism at the door and enter the future without illusions. . . . Authentic hope . . . must be rooted in the truth as best we can see it.[9]

Cancer researcher Dr. Jerome Groopman, makes a similar point about authentic hope:

> Many of us confuse hope with optimism, a prevailing attitude that "things turn out for the best." But hope differs from optimism. Hope does not arise from being told to "think positively" or from hearing an overly rosy forecast. Hope, unlike optimism, is rooted in unalloyed reality.[10]

In other words, authentic hope is grounded in unflinching honesty. A story of environmental politics that fosters authentic hope must be grounded in a clear-eyed look at planetary realities. Like a patient needing an honest and complete medical report, people need to confront the distressing realities of our planetary condition.

More than seven decades ago, conservationist Aldo Leopold penned the powerful phrase "a world of wounds."[11] Since then, the scars and ongoing injuries of ecological degradation, pollution, and climate change have relentlessly worsened our world of wounds. Treasured landscapes have been paved over for subdivisions, parking lots, and strip malls; countless ecosystems are degrading, burning, and flooding due to global warming-fueled storms and wildfires. Sledding, skiing, skating, and other winter sports are endangered by warmer winters and loss of snow. Popular activities such as hiking, gardening, camping, fishing, and even Little League baseball are compromised by dangerous heat and humidity. High school athletes' health and performance are threatened by heat and air pollution. No one is immune.

Living in a world of wounds has emotional repercussions for all of us. People of all ages experience depression and anxiety exacerbated by fears of future environmental losses. Environmental scientists experience ecological grief as they grapple with the implications of their own findings. A sustainability scientist wrote, "Being witness to the demise or death of what we love has started to look an awful lot like the job description."[12] A marine biologist who studies coral reefs admitted, "We come back from our field seasons increasingly broken."[13] An entire generation of young people is coming of age in a world of wounds. In 2021, a large, international study published in *The Lancet* examined children's and young people's feelings about climate change and government responses to climate change.[14] Reporting on the study, *National Geographic* noted that almost half of the 10,000 respondents (ages sixteen to twenty-five) reported that concerns about the state of the planet were interfering "with their sleep, their ability to study, to play, and to have fun."[15]

Our world of wounds has led to the emergence of new concepts and terms to capture the emotional repercussions. Terms such as "solastalgia" (the distress produced by environmental change), "eco-grief," "loss of ontological security" (fear that the continuity of life is threatened), and "eco-anxiety" (a chronic fear of environmental doom) have emerged to describe the emotional and existential distress associated with living in a world of relentless, ongoing environmental damage, irreversible ecological losses, and climate change. For many, solastalgia and eco-anxiety are underlain by love of the natural environment. Ecological grief researcher Ashlee Cunsolo points out, "We only grieve what we love. We are only scared to lose what matters."[16] Chris Jordan, a photographer known for his digital recreations of mass consumerism, was deeply shaken by his personal encounters with the devastating impacts of plastic pollution on Midway Island's Albatross population.[17] Pondering his experience of ecological grief, he wrote, "Grief is a felt experience of love for something lost or that we are losing. That is an incredibly powerful doorway. I think we all carry that abiding ocean of love for the miracle of our world."[18]

Countless Americans, young and old, are affected by eco-anxiety and eco-grief. A Pew Research Center survey found 70 percent of Americans feel "sad about what is happening to earth";[19] in 2021, the American Psychological Association found that more than two-thirds of adults surveyed reported experiencing some degree of eco-anxiety.[20] The Blue Shield of California NextGen Climate Survey found that eight out of ten young Americans are concerned about the health of the planet; two-thirds of young Americans ages eighteen to twenty-three are worried about the mental health impacts of climate change.[21] Amidst the cacophony of dire scientific predictions and the lived experiences of environmental loss, people of all ages need tools to cope with the distressing emotional realities of environmental decline. Like a punch in the gut, living through and learning about climate change, biodiversity loss, and a whole host of other environmental issues carries an emotional wallop.

Most people experiencing eco-anxiety and eco-grief suffer alone in our world of wounds, often wondering why they feel the way they do. Others are coping more publicly. To deal with her own eco-anxiety, in 2014, Brown University professor Kate Schapira set up a booth in Providence's city parks modeled after Lucy's booth in the *Peanuts* cartoon. Schapira's sign "Climate Anxiety Counseling" invited people to sit down and chat about global warming. Surprised by the number of people eager to talk about their global warming worries, Schapira continued the project for a number of years; her book based on the project was released in 2024.[22] Others are convening funerals for shared mourning over lost places. In 2019, a funeral to commemorate the loss of the Okjökull glacier in Norway—the first glacier lost as a result of global warming—received worldwide media attention.[23] Similarly, in 2020, the Oregon Glacier Institute held a funeral vigil for the "death" of Oregon's Clark Glacier; it had melted completely away.[24]

If authentic hope is rooted in "the truth as best we can see it," there is more to the equation than facing the realities of planetary, ecological decline. People also need to confront the limitations and distortions of the prevailing story of environmental politics. The existing story compounds our world of wounds by heaping on political wounds. If we are to rebuild a bipartisan environmental movement, we also must confront our world of political wounds. The usual story of bipartisan political failures compounds ecological grief and anxiety and drains people of hope. Typical accounts ignore the actual story of environmental politics: American conservatives' long-running attack on the environmental movement. If authentic hope is grounded in "unalloyed reality," then members of the public need to understand that conservatives have been waging a decades-long, systematic assault on the environmental movement, environmentalists, and the science that undergirds the environmental movement.

For more than thirty years, conservatives in government, non-profit think tanks such as the Competitive Enterprise Institute, the Cato Institute, the Heartland Institute, and others; radio talk show hosts such as Glenn Beck and the late Rush Limbaugh; media outlets such as Fox News, the Sinclair Broadcast Group, and Breitbart; and free-market Christian evangelical groups such as the Cornwall Alliance have politicized the environmental movement and environmental science. With a slew of slurs and stereotypes, they have transformed the environmental movement into a left-wing radical movement to discredit it in the eyes of the public and undermine its power as an engine of social and political change. Unlike other social movements, the environmental movement is grounded in science. Challenging environmental science undercuts positions advocated by the environmental movement and undermines regulatory solutions. Conservatives have repeatedly challenged the science on a host of major environmental issues including acid rain, ozone loss, deforestation, biodiversity loss, and climate change. More recently, extinction denial is creeping into public discourse.

The triumph of conservatives' efforts is most vivid in the realm of climate change politics. Conservatives have relentlessly challenged the reality of global warming and the legitimacy of climate change science to manufacture doubt and confuse the public. By equating climate science with policy, particularly policy positions advocated by "left-wing environmental extremists," conservatives have cemented the partisan divide. By characterizing climate science as a political agenda advanced by "alarmist scientists" and liberals, conservatives have undercut the authority of environmental science, weakened the scientific case for more stringent legislation, and further undermined the environmental movement. Their efforts have paid off. In 2017, a speaker at the Conservative Political Action Conference described environmentalists as "some of the worst people in the world."[25] Conservatives knew support for environmentalism would dwindle if they could convince the public that the environmental movement was a social movement supported by terrible people making left-wing political arguments masquerading as science.

ENVIRONMENTAL POLITICS AND PARADIGMS

The story of environmental politics goes deeper than the political manipulation of partisanship and science. Public understanding of the deeper story of environmental politics is constrained by analyses focused on the staples of political analysis: political actors, ideology, economics, and institutions. Missing from typical accounts of environmental politics is the earth. By ignoring the earth, we have missed the underlying story of environmental politics, a story about competing worldviews also known as social paradigms. Beginning in the 1970s, shortly after the birth of the modern environmental movement, researchers studying environmental values and attitudes concluded that a new environmental paradigm or worldview driven by people's environmental concerns had emerged to challenge the dominant worldview of industrial societies. The pressure exerted by this new environmental worldview—fueled by public concern and the growing body of environmental science—is threatening to conservative defenders of the dominant social paradigm, a worldview wildly out of sync with planetary realities. Politicizing and discrediting the environmental movement and manipulating and politicizing environmental science became central tactics to protect the status quo.

In recent years, conservatives' science denial has gained some public prominence. The usual remedy for scientific misinformation is better education. However, scientific literacy alone will not solve the problem. Many ardent climate change deniers are among the most educated people on the planet. Science is important, but science is not enough. The real problem lies with the subterranean influence of our societal worldview. The dominant worldview in the United States and in most advanced, industrialized societies is grounded in human exemptionalism. According to this worldview, humans are separate and above nature and unencumbered by ecological laws. This dominant worldview minimizes the severity of environmental damage and reassures people that human ingenuity and technology can fix any problem we create.

If conservative secular or religious defenders of the dominant worldview have persuaded members of the public that the earth can withstand all human damage, then no amount of scientific information to the contrary will be persuasive. People reject information that does not align with their worldview. And if people have been persuaded that the environmental movement is simply a liberal political agenda out of step with ordinary Americans, then people have even more reasons to reject environmentalists and the science that undergirds the environmental movement. However, in the face of escalating planetary emergencies—with biodiversity loss threatening the ecological fabric of life and global warming accelerating and nearing dangerous tipping points—it is essential to tell this subterranean story of clashing worldviews and their defenders in environmental politics.

Unfortunately, much of the scholarship in the social sciences and humanities reinforces the dominant worldview. Like the broader society, these disciplines have been shaped by the intellectual tradition of dualism. For many social scientists, "the world consists only of humans engaging with humans, with nature no more than a passive backdrop."[26] Unmoored from the earth, most social science scholarship separates people from nature and ignores the crescendo of scientists' warnings about planetary dangers. As Clive Hamilton notes, "Our best scientists tell us insistently that a calamity is unfolding, that the life-support systems of Earth are being damaged in ways that threaten our survival. Yet in the face of these facts, we carry on as usual."[27]

There are some promising trends in recent social science scholarship. Social scientists known as "new materialists" center the material reality of the earth in their scholarly works.[28] A new generation of political scientists primarily from the subfields of political theory and international relations also are beginning to change the story, calling for a "re-materialisation of politics and political theory."[29] Beginning in the early 1990s, green political theorists began calling for a "re-materialised political theory that would no longer externalize nature from its conceptual construction."[30] Environmental political theorists such as Andrew Dobson, Robyn Eckersley, John Meyer, and others have produced a substantial body of work.[31] This scholarly trajectory has gained traction with the onset of the Anthropocene. Advocating a new "Planet Politics," Anthony Burke et al. assert that international relations (IR) scholars have failed to engage with "the planetary real."[32]

Some scholars in this new materialist tradition have declared the demise of IR as a discipline, "undone by the reality of the planet."[33] As Nicholas Hedlund-de Witt argues, "in light of intractable global crises such as climate change, it does not seem all that far-fetched to suppose that the world itself may be reasserting the reality principle."[34] Calling for "a new perspective in political science," Frank Biermann argues that "The classification of a new epoch in planetary history . . . is fundamentally changing how we understand our political systems."[35] This new class of political science scholarship grounded in the earth challenges longstanding ontological and epistemological assumptions central to the discipline. As Cameron Harrington asserts, "earth-system changes wrought by human action require the discipline to demystify its own ontological, epistemological, and methodological approaches."[36] Following the lead of these scholars, we need to "re-materialize" environmental politics.

A new story of environmental politics also must acknowledge the singular place of the environment in our politics. According to Meyer, "The mainstream political view of 'the environment,' at least in contemporary Western liberal-democratic societies, is that it is an issue area."[37] We can no longer treat the environment as simply one "issue area" among many others in American politics. There are unique characteristics of environmental issues that demand special attention. Many environmental policies affect the functioning of ecological systems and often have irreversible consequences for all future generations. Irreversibility contravenes the democratic norm of policy reversibility. When we blast the tops off Appalachian Mountains to extract coal, ancient landforms, streams, and complex ecosystems are obliterated forever. A new Congress or presidential administration cannot reverse the damage. Increasingly, U.S. domestic environmental policy decisions have irreversible effects with planetary consequences. As Naomi Oreskes notes, "In the past, human actions tended to be local and reversible, but increasingly our actions appear to be global and irreversible."[38] Irreversibility underscores the high stakes associated with many environmental policy decisions: there are no second chances, no opportunities to correct our mistakes. In environmental politics, the fate of future generations looms large.

ENVIRONMENTAL POLITICS AND SCIENCE

A new story of environmental politics demands that we ground environmental politics in the natural sciences. However, in a world of declining trust in science paired with partisan manipulation of science, how are members of the public to know if environmental science is credible? How do we know if we can trust the vast body of environmental science that literally shouts a warning to humanity? World-renowned science historian Oreskes argues that we should trust natural scientists because of their "sustained engagement with the world."[39] Oreskes reminds us of the easily overlooked fact that natural scientists are the experts who spend decades studying the natural world. In addition, science historians, science studies scholars, and social constructionists have critically analyzed the scientific method and the social production of scientific knowledge in the natural sciences. Their work offers important insights about the legitimacy and authority of environmental science of great relevance to environmental politics.

Oreskes offers cogent advice in her 2019 book, *Why Trust Science?* Emphasizing that trust is not the same thing as faith, Oreskes asserts that understanding the social production of scientific knowledge is key to the legitimacy of scientific findings.[40] Similarly, Myanna Lahsen maintains that constructionist analyses of climate science are essential to protecting the legitimacy of climate science. Referencing her own studies of climate scientists and "politicized anti-environmental scientists," she notes that constructionist analyses of how the science produced by both mainstream climate scientists and anti-environmental scientists is affected by social, psychological, and political factors revealed key differences that enhanced the legitimacy of climate scientists and their work.[41] In other words, Lahsen argues that "examining the socio-political aspects of climate science is therefore the best way to protect the integrity—and, thus, the authority—of science."[42] Examining the social production of scientific knowledge highlights the fact that scientists do not work in isolation. Warnings of planetary peril do not come from a lone genius or a select group of scientific mavericks. Individual scientists who devote their professional lives to studying the natural world are part of large, diverse communities of scientists who regularly and formally interrogate and challenge each other's work. In other words, as Oreskes points out, "science is fundamentally consensual."[43]

Scientific consensus is central in the environmental sciences. Major scientific assessments of global issues such as climate change, acid rain, and biodiversity loss are based on thousands of studies in the relevant scientific literature. One of the hallmarks of these large scientific assessments is consensus-based decision-making. Consensus allows diverse groups of scientists to speak with one voice in important deliberations informing public policy. For example, since its inception in 1988, the IPCC, the United Nations body charged with conducting comprehensive assessments of climate change science—and the social and economic impacts of climate change—has relied upon consensus in the development of its assessments and reports. In a similar fashion, the Intergovernmental Science-Policy Platform on Biodiversity and Ecosystem Services (IPBES), an independent intergovernmental body modeled after the IPCC to assess the scientific evidence surrounding biodiversity loss, also relies upon scientific consensus in their assessments and reports to policymakers.

The scientific consensus on anthropogenic global warming (AGW) has been well-studied. In 2004, Oreskes was the first to evaluate the strength of the scientific consensus on AGW. Oreskes examined 928 abstracts published in refereed scientific journals between 1993 and 2003. She found that none of the papers disagreed with the scientific consensus on AGW. In her article "The Scientific Consensus on Climate Change," Oreskes acknowledged that the scientific consensus might be wrong. However, she offered a cautionary note:

> Our grandchildren will surely blame us if they find that we understood the reality of anthropogenic climate change and failed to do anything about it. . . . There is a scientific consensus on the reality of anthropogenic climate change. Climate scientists have repeatedly tried to make this clear. It is time for the rest of us to listen.[44]

In 2013, John Cook et al. surveyed 11,944 peer-reviewed scientific articles on climate change published from 1991-2011. They found that 97 percent of climate scientists endorsed the scientific consensus on AGW.[45] In 2019, James Powell found that the scientific consensus on AGW is now 100 percent.[46] In other words, all credible climate scientists agree that humans are responsible for the dangerous overheating of the globe. Similarly, the IPCC's 2021 Sixth Assessment Report affirmed that humans are responsible for warming the atmosphere, oceans, and land.[47] Although the scientific consensus on biodiversity loss is well established, the consensus among biologists has not been studied like the consensus on AGW. Biodiversity loss is less visible to the public and not due to identifiable industrial actors in the same way that fossil fuel extraction and production fuel global warming. Biodiversity loss also receives less media coverage and has not yet been politicized to the same degree as climate change.

Science studies scholars have studied the implications of consensus-based decision-making in major scientific assessments. Although conservatives have consistently characterized environmental scientists as "alarmist," research reveals that climate scientists are far from that. In fact, quite the opposite is true. Climate scientists working in teams that base their findings on scientific consensus tend to *underestimate* the dangers of climate change. In an article subtitled "Erring on the Side of Least Drama," Keynyn Brysse et al. found that scientists working on IPCC's climate change assessments exhibit scientists' typical conservatism. Guided by the traditional scientific values of rationality, objectivity, and self-restraint, they argue that scientists demand more robust evidence in support of "surprising, dramatic, or alarming conclusions" than evidence for less alarming findings in line with the scientific status quo.[48] In other words, scientists involved in consensus-based assessments typically dismiss credible findings deemed scientific outliers. The result of this scientific conservatism is that scientists participating in large assessments of the scientific evidence of climate change "may have underestimated the magnitude and rate of expected impacts of anthropogenic climate change."[49]

The story of glaciologist Jason Box illustrates how scientists may dismiss conflicting views that fall outside of the scientific consensus. Box has installed and maintained a network of more than twenty automatic weather stations on Greenland's inland ice over the course of nineteen expeditions to Greenland. Although Box has published multiple peer-reviewed scientific studies and was a contributing author to the 2007 IPCC Report that was awarded the Nobel Peace Prize, he often has been regarded as a maverick by others in the scientific community.[50] In 2012, Box predicted that within a decade, the entire surface of the Greenland ice sheet would melt. His prediction was dismissed as "alarmist claptrap" by scientists who adhered to the scientific consensus. A few months after Box's prediction, much of Greenland's surface ice and snow melted within just one week.[51] Box's prediction about the pace of glacial melt in Greenland was accurate; if anything, "the maverick" underestimated the speed of surface melting in Greenland. In other words, glaciologists focused on the scientific consensus about the dynamics and pace of the melting of large ice sheets like Greenland dismissed Box's prediction as an outlier; it fell outside of the scientific consensus. His story also illustrates how the scientific consensus on the speed of polar ice melt underestimated the actual pace of global warming and melting in Greenland.

Building on Brysse et al.'s earlier work, an internationally renowned research team probing the specific role and consequences of consensus-based decision-making in major scientific assessments of acid precipitation, ozone depletion, and global warming-driven sea level rise found that the process of building consensus reinforces scientific conservatism with a bias toward "reassuring" rather than "alarming" conclusions. Their findings are sobering: "If consensus is viewed as a requirement, then scientists may avoid areas of controversy—even when these are recognized as scientifically significant—and underestimate potentially severe effects."[52] In other words, in the face of accelerating global warming—what the IPCC calls a "code red for humanity"—consensus-based scientific assessments alerting the world to the dangers of climate change have in all likelihood *underestimated* the speed of global warming and the severity of the threats. Speaking to a reporter for *Esquire* in 2015, renowned climate scientist Michael Mann confessed that about once a year, he has "nightmares of earth becoming a very alien planet."[53] In November 2019, 11,000 scientists released a letter with yet another warning stating "clearly and unequivocally . . . planet Earth is facing a climate emergency."[54] Knowing that this statement of planetary emergency is based on consensus science that systematically *underestimates* the pace and dangers of global warming should give us all nightmares.

Given the crescendo of scientists' warnings, the challenges before us in both education and politics are clear. According to a 2021 UN report, "The well-being of today's youth and future generations depends on an urgent and clear break with current trends of environmental decline."[55] We must heed the measured warnings of environmental scientists who "err on the side of least drama." It is time to act to protect the biosphere and a stable, livable climate. The well-being of today's youth—and all the rest of us— also depends on a clear break from the typical story of environmental politics. We must reclaim and rebuild a bipartisan environmental movement.

Together we can weave a new story of environmental politics.

CHAPTER OVERVIEW

Chapter 1 provides the conceptual framework for understanding the under-lying forces at work in environmental politics: the dynamics of competing worldviews or social paradigms. Conventional analyses of environmental politics ignore this powerful explanatory framework that transcends par-tisanship. Environmental politics is best understood as a contest between a relatively recent societal worldview referred to as the new ecological par-adigm (NEP, which was originally termed the "new environmental par-adigm") and the traditional societal worldview known as the dominant social paradigm (DSP).[56] The alternative ecological worldview is congru-ent with environmental science and the scale of our planetary emergency. Grounded in human exemptionalism, the dominant worldview is totally at odds with ecological realities. The environmental movement represents the public face of the new alternative worldview. The voluminous body of empirical environmental science paired with the power of a popular environmental movement poses a threat to conservative defenders of the dominant worldview. Politicizing, distorting, and delegitimizing the en-vironmental movement and environmental science have become central tactics used by conservative defenders of the dominant worldview. For too long, we have ignored the subterranean role of clashing paradigms and their defenders in environmental politics.

Chapter 2 provides necessary background on the state of the earth in the proposed new geological epoch known as the Anthropocene. If we are to focus environmental politics on the imperatives of planetary protec-tion, we need to understand the planet we seek to protect. Understanding the challenges of the Anthropocene also is important in order to compre-hend the stakes in environmental politics. Characterizing the Anthropo-cene as the Age of Humans invokes compelling visions of human control of nature. However, the Age of Humans represents a fundamental misunder-standing of nature and the behavior of the Earth System in the proposed epoch. Believing we can control nature is grounded in flawed assumptions embedded in society's dominant worldview and in misperceptions about the Anthropocene. In the Anthropocene, the Earth System is an increas-ingly unpredictable, active player in human affairs. In our journey into the Anthropocene, we are sailing into an era of planetary destabilization: social, economic, political, and ecological stability are at stake. Thus, the fate of future generations looms large in the Anthropocene. The chapter concludes by examining irreversible, ecological losses in the Anthropocene and the implications for democratic governance and intergenerational equity.

Chapter 3 begins the story of conservatives' campaign to undermine the political power and momentum of the environmental movement. For decades, conservatives in government, media, think tanks, and evangelical religious groups have been engaged in an assault on the environmental movement running on parallel tracks. Chapter 3 relates how prominent conservatives and their allies have waged a war of stereotypes and allegations about environmentalists that continues to this day. Conservative politicians, pundits, and radio hosts have caricaturized environmentalists with a slew of slurs and derogatory labels that have evolved from the relatively innocuous "tree huggers" to "eco-nazis," "jihadists," and a "native evil." Conservatives also have politicized the movement by aligning environmentalism with a liberal agenda: the accusations range from calling environmentalists "socialist watermelons"—green on the outside, red on the inside—to calling environmentalism a "woke" social justice movement. This chapter provides an overview of conservatives' attacks on environmentalists and explores the implications of discrediting the environmental movement as a driver of public opinion and political change.

Chapter 4 continues the story of conservatives' efforts to disempower the environmental movement by undermining the scientific evidence that undergirds the movement and motivates supporters. Soon after the first Earth Day, conservatives recognized the twin threat posed by the environmental movement. Unlike most other social movements, environmentalism relies heavily on scientific evidence. The combination of a popular bipartisan social movement supported by a growing body of environmental science could be a powerful catalyst for social and political change. Inspired by the original "merchants of doubt" who worked to discredit the health effects of tobacco, they trained their sights on environmental science and scientists.[57] Challenging the legitimacy of environmental science became an important weapon to undermine environmentalism and block environmental legislation. The chapter provides a chronological overview of the siege on environmental science, including the climate change denial campaign, and concludes with the next chapter of environmental science denial, extinction denial.

Chapter 5 asks, how did the conservative assault on environmental science become so deeply embedded in our politics? Enveloped within the politicization of science in environmental politics is a neglected story that begins in the halls of academe with the emergence of a school of thought known as social constructionism. Oversimplified notions of social constructionism gave conservatives a tool to politicize and delegitimize environmental science. The chapter traces the emergence of climate science critiques among academic constructionists and the cross fertilization between constructionist scholars and conservative think tanks known for manufacturing distortions of environmental and climate science. Through the lens of constructionism, environmental science is characterized as merely one viewpoint among other legitimate perspectives on environmental issues; the scientific consensus on climate change and other major planetary threats becomes merely a political consensus. The chapter concludes by examining constructionist analyses of the Anthropocene to illustrate the influence of social constructionism on interpretations of environmental science in scholarship and education.

Chapter 6 deconstructs scientific uncertainty in environmental politics and policymaking. Framing science through the lens of social constructionism elevates the role of scientific uncertainty and downplays the significance of the scientific consensus on our major planetary threats. Highlighting scientific uncertainty plays into the hands of the conservatives who have weaponized and manufactured scientific uncertainty to erode public concern and stymie political action. Thus, examining scientific uncertainty in environmental politics requires a nuanced treatment that differentiates between legitimate and manufactured scientific uncertainty. However, making these important distinctions can pose pedagogical challenges and professional risks for college and university instructors and textbook authors. The chapter concludes by examining the tensions inherent in science-based policymaking in the context of democratic norms. The chapter offers a selected review of scholarship on innovative policy processes that protect both scientific integrity and democratic deliberation.

Chapter 7 concludes the story of conservatives' denial of the gravity and implications of global warming and examines some of the consequences of their intransigence. After thirty-five years of climate change denial, conservatives have embraced a suite of risky technologies, known as geoengineering, for managing the global climate to continue our reliance on fossil fuels without openly acknowledging the reality of human-caused global warming. Geoengineering fits squarely within the dominant societal worldview, a worldview that prioritizes technological solutions to environmental problems while simultaneously downplaying the severity of environmental problems. The chapter provides an overview of geoengineering technologies and the intertwined story of geoengineering and climate change politics. The chapter concludes by examining the implications of relying on geoengineering to perpetuate an economy driven by fossil fuels grounded in a worldview dangerously out of touch with planetary realities.

Chapter 8 turns to the question of how to counter the myths of the dominant societal worldview and bridge the partisan divide. The chapter probes cognitive and emotional barriers including manufactured partisanship and cognitive biases as well as opportunities for change. The chapter examines how people can begin a journey away from the ecological delusions of the dominant paradigm to a worldview grounded in planetary realities. Integrating climate change communication research and dispute resolution theory and practice, the chapter explores ways to communicate about climate change, biodiversity loss, and other contested environmental issues by entering conversations through the "back door" of interests. Instead of challenging entrenched positions head-on, if we discuss widely shared concerns about public lands, wildlife conservation, sustainability, plastic pollution, public health, and the many other collateral damages of our reliance on fossil fuels, we can engage in productive conversations that sidestep environmental science denial, breach the walls of manufactured partisanship, and dislodge the roots of the dominant worldview. The chapter offers pathways for rebuilding a bipartisan environmental movement essential for political change and long overdue environmental progress. The chapter concludes by considering whether rebuilding a bipartisan environmental movement could have broader implications for American society and politics.

Chapter 9, the final chapter, returns to the essential role of hope in our "world of wounds." The authentic hope that we need to overcome despair and mobilize politically to address ecological decline and climate change must be based on the truth about our planetary predicament *and* environmental politics. Revisiting the story of paradigms and partisanship, I argue that telling the authentic story of environmental politics means we must challenge the false equivalencies that distort the reality of environmental politics and cause despair. We must be willing to confront the political actors who have politicized the environmental movement and science, manipulating and falsifying the scientific realities of climate change, bio-diversity loss, and other major planetary threats. Frank discussion about the realities of environmental politics enables us to harness the power of emotion in our politics by unleashing the productive energy of anger. The combination of scientific and political realism, anger, and hope can overcome eco-anxiety and political paralysis. Ironically, the potent blend of realism, anger, and hope offers a pathway to true bipartisanship. The chapter argues that it is not too late to rebuild a bipartisan environmental movement grounded in Americans' shared love of our natural heritage.

ONE

ENVIRONMENTALISM AND WORLDVIEWS

Imagine that you've just arrived on the coast of Oregon to visit the old-growth forests. Perhaps you plan to visit Cape Meares State Park to see the oldest Sitka Spruce in the state. Towering thirteen stories high, you can only imagine how dwarfed you will feel at the base of the 800-year-old giant. As an avid angler, you were interested to learn that intact old-growth forests in Oregon and elsewhere in the Pacific Northwest anchor topsoil and protect the rivers and streams that will flicker with dancing light as salmon make their way upstream to spawn in late summer and fall. Before your trip, you read Suzanne Simard's *Finding the Mother Tree*.[1] How fascinating to learn that the ancient "mother trees" communicate with and support their own offspring! How touching to learn that as the mammoth mother trees reach the end of their lives, they transfer their storehouse of genetic wisdom to their seedling kin. What wisdom and insights will the towering mother trees share with you?

Once you arrive—perhaps you've stopped to get gas—you happen to overhear two foresters talking about plans to log old-growth timber. What? You hear them evaluating Oregon's old-growth trees in terms of board feet of lumber. You hear the foresters describing the ancient trees as "senescent." What does that even mean? The foresters are describing the venerable giants as economic commodities declining in value as long as they continue to grow. In the name of economic efficiency, the foresters want to cut them down. The foresters' vision of the old-growth forests is jarring and totally contradictory to your view of the forests before you've even had to chance to hike in the cathedral forests, as so many have described them. You cannot fathom why these foresters think we should cut more of the last vestiges of old-growth forests in the United States. In this puzzling encounter, you have just bumped into the reality of clashing worldviews. The foresters are looking at the forests through a different lens, as though they were wearing special glasses that focused their perceptions of the old-growth forests in an understandable way for them. The foresters' perceptions of the forests are foreign and incomprehensible to you because you and the foresters have different worldviews.

This chapter introduces a conceptual framework grounded in sociological research for understanding the underlying dynamics of environmental politics. The story of environmental politics is a story of competing paradigms, not simply competing partisans. It is a story about the contest between an environmental movement driven by an alternative worldview congruent with environmental science and the deteriorating condition of our planetary home versus defenders of the status quo operating under the influence of a worldview divorced from the planetary real. The environmental movement and environmental science play a central role in this drama. Environmentalists and citizens worried about the environmental deterioration they see and experience in our world of wounds often look to science to explain the problems. An enormous body of environmental science tells us that business as usual is not sustainable. Industrial civilization is wreaking havoc with the ecological fabric of life and the very integrity and functioning of planetary systems. This contest between defenders of the status quo and environmental advocates of a competing, earth-centered worldview drives environmental politics.

INTRODUCTION TO PARADIGMS

How do we understand the world around us? Do we simply take in raw data about the social and physical world? How do we make sense of our perceptions? Our perceptions are filtered through worldviews or social paradigms. Worldviews help us interpret reality. Like putting on a special pair of glasses, paradigms provide a lens for perceiving and interpreting complex social realities. Social paradigms provide a belief structure that helps people make sense of the world. In other words, social paradigms offer a coherent story of how the world works.

Every society has a dominant social paradigm comprised of the values, beliefs, habits, and institutions that, taken together, help citizens interpret the complexity of the social world. A paradigm is not dominant in the sense that it is embraced by the majority of people in society. Rather, a paradigm is dominant in that it is embraced and protected by powerful society elites. The dominant social paradigm legitimizes and maintains the institutions and practices of society; conversely, the norms, institutions, and practices of society reinforce the dominant paradigm. In many developed nations including the United States, the dominant social paradigm evolved from an industrial worldview. Priorities associated with the dominant paradigm include individualism, a free-market economy, unlimited economic growth, and maximizing wealth. Highly anthropocentric, this dominant worldview is based on the Judeo-Christian view of human progress and mastery over nature.

Although the dominant paradigm is embraced by society elites, it is not universally adopted by all members of society. Society can contain alternative social paradigms that contrast with traditional, common ways of thinking. The notion of an alternative paradigm that contrasts with the dominant paradigm is foundational to understanding the dynamics of environmental conflict and environmental politics. Researchers in sociology and political science found that not long after the birth of the modern environmental movement, a new environmental paradigm (NEP) or worldview emerged to challenge the dominant social paradigm (DSP) in the United States and in other developed nations. Beliefs and priorities associated with the new environmental paradigm include protecting nature to avoid environmental catastrophe, careful planning to avoid environmental risks, and new forms of political and economic organization to stop the massive destruction of the environment sanctioned by the DSP. The biocentric NEP does not elevate humans above the rest of nature; other organisms are seen as having inherent worth independent of human utilitarian needs.

Paradigms do not simply help us interpret social realities. Paradigms also shape our understanding of physical reality. The NEP and DSP comprise profoundly different views on the nature of our physical environment, the earth. In contrast to the overwhelming body of scientific evidence of ecological decline, the DSP is grounded in the belief that the earth was created to benefit humanity. Assumptions embedded within the DSP include the notion that the earth is an unlimited storehouse of resources created for human use and that humans have the right to exploit nature.

Environmental damage is justified by the unwavering conviction that the earth can withstand any harm humans inflict in pursuit of economic progress; any problems we cause can be corrected by human ingenuity and advanced technology. In contrast, the NEP is based on a view of the earth as finite and valuable for its own sake, and seriously jeopardized by human impacts. The NEP worldview recognizes that nature is a dynamic force that humans cannot control. These divergent views of the earth—worldviews in the most literal sense—arise from contrasting assumptions about humans' place in nature. Grounded in human exemptionalism, the DSP assumes that humans are unfettered by ecological constraints. Conversely, the NEP recognizes that humans cannot escape the same ecological laws and principles that regulate the growth, development, and well-being of all other species. We are not exempt from ecological limits and the laws of physics.

THE NEW ENVIRONMENTAL PARADIGM AND THE EARTH

Evidence for the emergence of the alternative environmental paradigm challenging the DSP is based on a large number of studies utilizing a measure called the New Environmental Paradigm Scale first developed in 1978. Sociologists and political scientists found evidence of an alternative environmental paradigm among members of the American public soon after the birth of the U.S. environmental movement in the late 1960s. Over the years, researchers from diverse disciplines have found continued evidence of the existence of the NEP in society in the United States and elsewhere.[2] Numerous studies have found the NEP Scale to be predictive of a variety of environmental attitudes and beliefs including opinions about the severity of environmental problems, renewable energy, climate change and global warming, and support for environmental policies. The power and relevance of the NEP Scale rests in part on the fact that it measures "primitive beliefs" about the nature of the earth and humans' relationship with the earth.[3] Although the NEP Scale is just one of many measures of environmental attitudes and concern, it has become the most extensively used measure of environmental concern in the world. It has been utilized in hundreds of studies in scores of nations in spite of some substantive and methodological criticisms.[4]

Some researchers have criticized the NEP Scale's inability to predict environmental behaviors,[5] its failure to adequately integrate environmental ethics[6] as well as humans' intrinsic and spiritual connections to nature,[7] and its inability to address deep green concerns.[8] The NEP Scale's statement about the fragility of nature's balance does not reflect contemporary ecological science; the concept of the "balance of nature" has been replaced with the understanding that dynamic change is the norm in natural systems. Some analysts have argued that the influence of the NEP Scale in research is waning, particularly in the United States. However, no alternative to the NEP Scale has been widely adopted. Jennifer Bernstein reports that the number of research citations that reference the "New Environmental Paradigm" and the "New Ecological Paradigm" continues to grow.[9]

The NEP Scale continues to be a widely used measure in part because it has been revised to remain relevant. Bernstein notes that the NEP Scale's "lack of reference to explicit issues kept the scale relevant as environmental problems evolved."[10] As environmental issues grew from the more local and pollution-related concerns of the early environmental movement to the more global issues that define our Anthropocene journey, the original NEP Scale was revised to correspond with changing environmental realities and evolving public concerns.[11] In 2000, Riley Dunlap et al. revised the NEP Scale due in part to the growing prominence of an ecological worldview, arguing that "we are in the midst of a fundamental reevaluation of the underlying worldview that has guided our relationship to the physical environment."[12] They renamed the NEP Scale the New Ecological Paradigm Scale (see Appendix) to better incorporate elements of the ecological worldview central to the NEP; they also renamed the NEP the new ecological paradigm. Other researchers have confirmed that the revised New Ecological Paradigm Scale measures an ecological worldview. According to C. Xiao et al.,

> An ecological worldview (as measured by the NEP) . . . entails viewing human societies as embedded in and dependent upon ecosystems. It accepts that ecological limits can constrain growth and that exceeding them may disrupt ecosystems (with negative consequences for humans and other species).[13]

THE NEP AND ENVIRONMENTAL SCIENCE

Evidence suggests that a more ecologically accurate worldview is now prevalent among many in the American public. This ecological worldview undergirds the environmental movement. However, it is important to em-

ENVIRONMENTALISM AND WORLDVIEWS

phasize that as designed, the NEP Scale measures people's beliefs about the world and nature. If the NEP only offered a way of understanding changing environmental values and attitudes, we would be left with the standard fare of politics. Public opinion is important in politics; yet as abundant evidence suggests, public opinion is not necessarily congruent with science. Thus, we need to examine how beliefs about nature associated with the NEP correspond with the physical reality of the natural world.

The assumptions about nature embedded in the new ecological paradigm are scientifically defensible. The NEP view of the earth is largely congruent with ecosystem science, and consistent with the scientific consensus on the severity of our environmental problems at all spatial scales. The NEP view of the earth recognizes biophysical limits and the real possibility of ecological catastrophe if we continue to pursue the present course driven by the DSP. The following paragraphs explore the relationship between the ecological worldview of the NEP and systems theory. Exploring this complex relationship illustrates the ways in which the NEP is grounded in ecological realities.

Adam Davis and Mirella Stroink offer a robust analysis of the congruence between the ecological worldview of the NEP and ecological science in their article, "The Relationship Between Systems Thinking and the New Ecological Paradigm." They found that the NEP "uniquely predicted an enhanced ability to think in systems."[14] Systems thinking is central to our understanding of ecological systems at all spatial scales, from the local level to the overall Earth System. Systems thinking draws attention to the dynamic interconnections among parts of the system. Years before our current understanding of systems theory, the famous naturalist John Muir wrote, "When we try to pick out anything by itself, we find it hitched to everything else in the Universe."[15] Without the benefit of scientific training, Muir glimpsed the complex, interconnected realities of natural systems.

At the simplest level, systems thinking recognizes that the whole is greater than the sum of the parts. We cannot understand the dynamics of ecological systems by focusing only on certain parts or features of the system. For example, we cannot understand the dynamics of forest ecosystems by only looking at the trees. Within forest ecosystems, the above and below ground components of the ecosystem are interconnected and interact through complex pathways. A vast network of underground mycorrhizal fungi connects tree species with nutrients and transports chemical signals that influence the health of the overall forest system.[16] For exam-

ple, chemical signals from one tree species in distress can be communi-
cated to other species via the underground fungal network; the recipients
of these chemical signals change their own biochemistry to protect themselves from insects and disease.

Traditional forestry practices are based on efficiency-driven models of
resource management shaped by the DSP view of the earth as simply a
warehouse of resources for human use. Mainstream forestry fails to recognize the power of these complex interactions within forest ecosystems.
For example, Simard's research found that when Canadian foresters eliminated birch trees growing alongside Douglas Fir, the Douglas Fir trees
unexpectedly declined.[17] Driven by an industrial model of forestry, the
foresters assumed the birch were competing with the Douglas Fir (the desirable commercial species) based on a linear, mechanistic calculus about
available sunlight, nutrients, and water. In fact, complex chemical interactions between the birch and the fir, mediated through beneficial mycorrhizal fungi, protected the young Douglas Fir trees from root disease. The fir
would not thrive without the birch, even though the faster-growing birch
shaded the Douglas Fir seedlings. The firs' unexpected decline could not be
corrected through predictable inputs of fertilizers and pesticides driven
by DSP thinking.

For many people, systems thinking requires a perceptual shift in how
we think about the world. Our understanding of the world has been pow-
erfully shaped by mechanistic patterns of thought associated with the
DSP. From the perspective of the DSP, we think we can manage and control natural systems as though they operate like machines, where change
is linear, predictable, and manageable. In contrast to this mechanistic
worldview, systems thinking reveals that natural systems are "complex
adaptive systems." Unlike machines, complex adaptive systems (CAS) are
self-organizing systems. The individual components of CAS are interconnected and interact with each other through complex mechanisms. In contrast, human-made machines can be *complicated,* but they are not *complex*
self-organizing systems. CAS change themselves independently of human
interventions; machines do not have the agency to change themselves. For
example, car engines are complicated; however, they are not complex systems. Automobile engineers can predict and change how car engines operate by modifying parts of the engine system. But the parts of the engine
do not re-arrange themselves in surprising ways. Similarly, we will never
walk into our kitchens to find that the microwave has transformed itself

into a conventional oven. In response to natural or human causes, self-organizing CAS change themselves in unpredictable ways that often come as a surprise.

Central assumptions of the NEP worldview are consistent with core features of complex adaptive systems and the growing understanding of the connections between humans and nature. Primary tenets of the NEP worldview that embody features of complex adaptive systems include "anti-exemptionalism, the reciprocal exchange between social, ecological, and economic systems, [and] the delusion of human dominion over the natural world."[18] Ecologists and resilience theorists tell us that humans live and operate in dynamic complex adaptive systems composed of linked social and ecological systems. Resource managers and many scholars now speak of social-ecological systems rather than human societies and natural ecosystems as separate spheres. Social-ecological systems are everywhere. Cities, towns, and farms are social-ecological systems. So are national parks, forests, and fisheries. As resilience theorists Brian Walker and David Salt explain, "Be it a farm, a business, a region, or an industry, we are all part of some system of humans and nature."[19] Humans and nature are inextricably interconnected.

The scientific understanding of the embedded relationship of humans and nature is consistent with one of the principal NEP beliefs: humans are part of nature. This NEP view of the human relationship to nature is consistent with the realities of social-ecological systems: social, economic, and ecological systems are interwoven and influence one another in significant ways. Both NEP adherents and systems thinkers acknowledge "system membership." NEP adherents and systems thinkers see themselves and all human beings "as components of complex adaptive social, economic, and ecological systems."[20]

In contrast, the separation of humans and nature is a central feature of the DSP. The fact that humans are part of linked social-ecological systems directly challenges the DSP conviction that humans are separate from nature and exempt from ecological constraints. Davis and Stroink found that the central NEP belief of anti-exemptionalism—the recognition that humans are not exempt from the laws of nature—"uniquely predicted a systems mindset."[21] Systems thinkers and adherents of the NEP both recognize that humans are part of, not separate from, natural systems. As part of natural systems, humans are not exempt from the ecological conditions, processes, and constraints that govern the behaviors and opportunities of all other species. This fundamental ecological reality is the antithesis of the exemptionalism that grounds the DSP and helps rationalize the unbridled faith in technology to overcome biophysical limits.

The recognition that humans are not exempt from ecological constraints also challenges the deep-rooted DSP tenet that humans can control nature. As Davis and Stroink point out, "Individuals possessing an ecologically conscious worldview also comprehend that human beings will never possess the ability to control and dominate nature."[22] The notion that we can control nature does not match the realities of how complex adaptive systems—including the Earth System—function. We cannot control the behavior of self-organizing complex adaptive systems. As a society driven by the DSP, enamored with the power of technology, we have never grasped the difference between power and control. The onset of the Anthropocene is evidence of human power; however, thinking that we can control nature demonstrates a willful and dangerous ignorance of ecological realities.

ENVIRONMENTAL POLITICS AND PARADIGMS

The role of clashing worldviews increasingly is being studied in the context of attitudes toward climate change and sustainability. Anne Pender argues that worldviews are a poorly understood and underutilized tool that could enable "more transformative governance responses to climate change."[23] Annick de Witt has examined the clash of worldviews in climate change politics and sustainable development.[24] Focusing on challenges in climate change adaptation, Shannon McNeeley and Heather Lazrus argue that perceptions of climate change risk are informed both by social interactions and "cultural worldviews comprising fundamental beliefs about society and nature."[25] Psychologists Gail Hochachka and Terri O'Fallon have utilized cognitive models of human development to examine individuals' evolving worldviews in relation to climate change.[26]

However, the subterranean influence of clashing worldviews is largely missing from most conventional analyses of environmental politics in the United States. Political theorist John Meyer examines the centrality of dichotomous worldviews in environmental thought.[27] Textbook authors Zachary Smith and Michael Kraft both introduce the concept of worldviews in environmental politics in their introductory chapters that examine public opinion and the role of competing values.[28] However, the dynamics of competing paradigms in environmental politics is not utilized as an explanatory lens in most Environmental policy scholarship. Most analyses of environmental policy and politics focus on political actors, values, ideology, institutions, economics, and the unique challenges of particular environmental policy issues.

For members of the public, the subterranean nature of paradigms makes them difficult to detect and easy to miss. We do not notice how the DSP shapes our culture, institutions, and politics. We overlook the fact that political ideologies are arguments used by political actors to justify policy choices and actions derived from the underlying worldview. Focusing on paradigms changes the story of environmental politics. The emergence of the NEP with its ecological worldview puts the earth, the dynamics of competing paradigms, the environmental movement, and environmental science at the center of environmental politics.

Years ago, political scientist Lester Milbrath anchored environmental politics in the earth; he argued that viewing contestation in environmental politics only in terms of the traditional left-right ideological spectrum is too simplistic and less relevant for beliefs about humans' relationship with nature.[29] Milbrath and several sociologists pointed out that paradigms transcend ideology. For example, Riley Dunlap and Kent Van Liere note that faith in continued resource abundance associated with the dominant paradigm is not associated with either liberalism or conservatism.[30] Similarly, Milbrath observed that political elites on the left and right both strongly endorse economic growth and material wealth. In other words, the body of work on the emergence of the NEP in society indicates that our study of environmental politics needs to go beyond traditional political variables. The story of environmental politics is not a simple story of competing partisans. It is a story of clashing paradigms.

When we examine our domestic environmental politics through the lens of competing paradigms, the primary axis of conflict becomes the clash between conservative defenders of the DSP, who prioritize material wealth, versus environmental advocates, who identify with the NEP (Figure 1.1). Milbrath described environmentalists—the public face of the NEP—as the "Vanguard"; he described defenders of the DSP as "Rearguard" Traditional Material Wealth Advocates. Milbrath put the general population of "Environmental Supporters" in a circle in the middle of the model (Figure 1.2). The circle of supporters demonstrates that public support for environmental protection not only transcends partisanship, it also crosses

FIGURE 1.1   *Model of Environmental Politics*

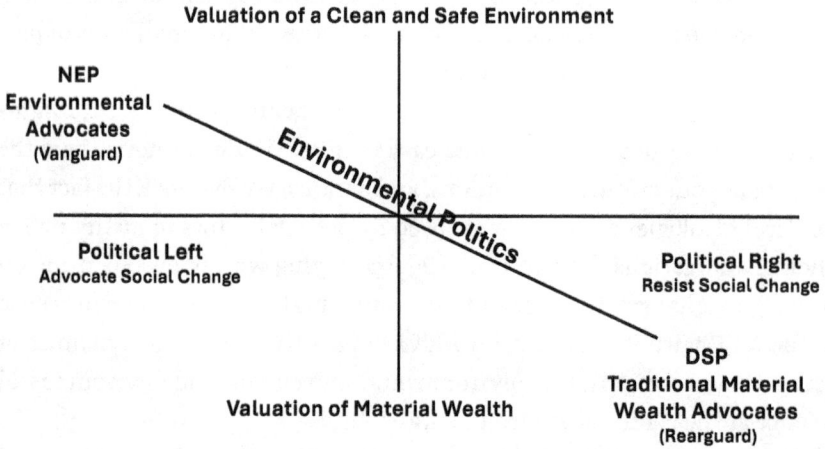

Valuation of a Clean and Safe Environment

NEP
Environmental
Advocates
(Vanguard)

Environmental Politics

Political Left
Advocate Social Change

Political Right
Resist Social Change

DSP
Traditional Material
Wealth Advocates
(Rearguard)

Valuation of Material Wealth

Source: Adapted from Lester W. Milbrath, *Environmentalists: Vanguard for a New Society*, 1984, p. 24. Copyright © 1985 by the State University of New York Press. Reprinted with permission.

FIGURE 1.2   *Model of Environmental Supporters*

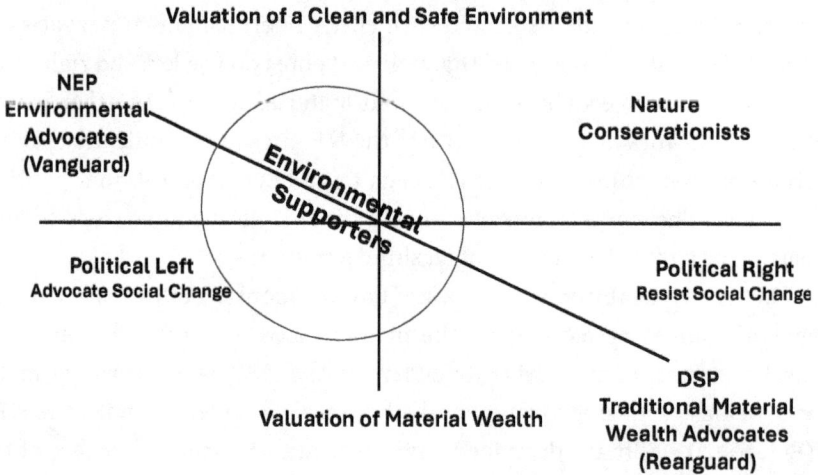

Valuation of a Clean and Safe Environment

NEP
Environmental
Advocates
(Vanguard)

Nature
Conservationists

Environmental Supporters

Political Left
Advocate Social Change

Political Right
Resist Social Change

DSP
Traditional Material
Wealth Advocates
(Rearguard)

Valuation of Material Wealth

Source: Adapted from Lester W. Milbrath, *Environmentalists: Vanguard for a New Society*, 1984, p. 24. Copyright © 1985 by the State University of New York Press. Reprinted with permission.

paradigms. One can prioritize material wealth while simultaneously valuing environmental protection. Similarly, some environmental supporters and traditional nature conservationists may not be strong proponents of social change.

When environmental advocates and rearguard defenders of the DSP clash in environmental politics, both sides often find the encounters infuriating and almost undecipherable. Proponents of conflicting worldviews are looking at the world through fundamentally different lenses, like they are literally viewing different worlds. Stephen Cotgrove characterized environmental politics as actors facing each other "in a spirit of exasperation, talking past each other with mutual incomprehension. It is a dialogue of the blind talking to the deaf."[31] This exercise in shared frustration is not caused simply by confronting political actors driven by different ideological positions or values. This "dialogue of the blind talking to the deaf" results when actors with fundamentally different worldviews interact.

What is reasonable and appropriate from the perspective of defenders of the DSP often is viewed as unthinkable and dangerous to proponents of the NEP, and vice versa. For example, defenders of the DSP often make political arguments and policy proposals grounded in the firm belief that the ecological crisis has been greatly exaggerated. This rearguard DSP belief is the antithesis of the NEP view—backed by consensus science—of the severity of our planetary problems. For example, when conservative politicians known as lukewarmers claim that climate change is real but not a serious problem, proponents of the NEP, keenly aware of the dangerous realities of climate change, shake their heads in dismay. Conversely, when NEP adherents argue that environmental protection must be prioritized over endless economic growth, defenders of the DSP view such claims as unthinkable nonsense.

We see this clash of paradigms vividly in conflicts over national forest policy. The first chief of the U.S. Forest Service, Gifford Pinchot, once said, "Forestry is tree farming. Forestry is handling trees so that one crop follows another."[32] During the battles over the old-growth forests of the Pacific Northwest in the 1980s and early 1990s, it was clear that foresters and commercial timber interests viewed forests through the lens of the DSP: they saw old-growth forests simply as timber resource—a crop—for humans to exploit. The foresters viewed the old-growth forests in terms of "board feet of lumber," losing economic value the longer they stood unharvested. In contrast, those viewing the old-growth forests through the lens of the NEP

saw the forests as "cathedral forests" full of ancient trees with intrinsic ecological, aesthetic, and spiritual value. They saw the ancient forests as biologically diverse forest ecosystems that protected watersheds and housed rare
and endangered species. They understood that clear-cutting the old trees
would irreversibly destroy the ancient forest communities.

The views of the NEP proponents defending the ancient forests were
consistent with the research on forest ecosystems and the role of the largest trees in forest communities. These trees—dubbed "mother trees" by
pioneering forestry researcher Suzanne Simard—anchor entire forest
ecosystems.[33] Through the complex underground network of mycorrhizal
fungi, mother trees communicate with and support their own offspring,
shuttling carbon to their seedlings as needed. As mother trees reach the
end of their lives, they transfer their genetic storehouse of wisdom about
adaptation and survivability to their kin. The Forest Service and commercial timber interests—viewing the forests through the industrial lens of
the DSP—could not have been more wrong about the forests. Seeing forests
as simply a storehouse of resources for humans to exploit, they cut the
huge, ancient mother trees and compromised the integrity and genetic resilience of the forest ecosystems.

ENVIRONMENTAL POLITICS AND PARADIGM SHIFTS

The NEP represents a significant threat to defenders of the DSP. The ecological worldview of the NEP fundamentally contradicts the DSP worldview.
Environmental studies and sustainability scholars have written extensively about the environmental, economic, and social changes necessary
to protect planetary systems and forge a sustainable society. As the public
face of the NEP, environmentalists advocate for fundamental changes in
social, political, and economic systems to safeguard the earth and future
generations' ecological inheritance. The pressure of the NEP raises the
specter of a paradigm shift in American society. In response, conservatives
have mounted a powerful rearguard effort to protect the DSP.

As ardent champions of the DSP, what exactly are conservatives defending? Many analyses of environmental politics focus on conservatives'
antipathy to regulation. Much of the success of the environmental movement in the United States is a story of the power of federal regulation to
correct negative externalities such as air and water pollution. Conservatives are staunchly opposed to the use of regulation as a policy tool. Con-

servatives view regulations as impediments to the operation of the free market, the antithesis of economic liberty. As Naomi Oreskes and Erik Conway emphasize, economic and political liberty are revered as the twin pillars of freedom that undergird conservatives' vision of the American way of life.[34] Beyond ideological opposition, federal regulations termed "command and control" are criticized by conservatives and regulated industries for being inflexible and too costly. Richard Andrews notes that regulated industries have proposed alternatives to command and control, including self-regulation, market trading of emission permits, and pollution prevention.[35] The 1990 Clean Air Act successfully incorporated market-based incentives for mitigating acid rain; and in 1990, Congress passed the Pollution Prevention Act.[36] However, self-regulation does not have a strong track record. When William Ruckelshaus, first administrator of the Environmental Protection Agency (EPA), was asked what he thought of the voluntary regulatory initiatives implemented during the Reagan administration, he remarked, "the only thing voluntary about voluntary regulation is if EPA voluntarily chooses not to enforce the law."[37]

Regulatory battles typically are at the center of environmental politics; regulatory agencies like the EPA write the rules necessary for policy implementation. Environmental rulemaking is highly detailed, scientifically and technologically complex, and sometimes protracted. For example, commenting on the implementing regulations for the 1990 Clean Air Act (CAA), Jacqueline Vaughn and Hanna Cortner observed that the legislative changes to the CAA "led to more than 10,0000 pages of implementing regulations requiring years of regulatory effort."[38] Compared to legislative deliberations, rulemaking usually receives little media attention.[39] Judith Layzer notes that conservatives have "capitalized on the relatively low visibility of bureaucratic . . . decision making to weaken existing policies or prevent adjustments to new or growing risks."[40] Legendary legislator U.S. Representative John Dingell once said, "I'll let you write the substance on a statute and you let me write the procedures [regulations], and I'll screw you every time."[41] Working within the "black box" of regulatory agencies such as the EPA, conservatives have successfully blocked, weakened, and delayed new regulations and advanced a compelling antiregulatory public narrative that has legitimized distrust of government, science, and environmentalists.[42] Conservatives' victories in the bureaucratic arena, paired with their success in stoking anti-regulatory sentiment in American society, raise additional questions about the motives driving conservatives' obstructionist, anti-regulatory agenda in environmental politics. We can add new dimensions to the story of conservatives' disdain for environmental regulation by factoring in the earth and the powerful role of worldviews in environmental politics.

Conservatives' ideological opposition to the proverbial red tape of bureaucratic regulation is fundamentally an argument to justify policy choices consistent with the underlying DSP worldview. By rejecting regulations, conservatives are protecting a worldview grounded in human exemptionalism that elevates humans above all else in the natural world and gives humans free license to exploit natural resources as a means of accruing wealth. In other words, conservatives are guarding more than economic liberty; they are protecting a comprehensive worldview along with the traditional values and societal practices justified and normalized by the dominant worldview.

For example, many conservatives perceive environmentalism as a threat to unrestricted private property rights, mass consumption, and the material comforts associated with the American way of life. Some conservatives have targeted the sustainability movement, described by Daniel Mazmanian and Michael Kraft as the "third epoch" of the environmental movement.[43] Sustainability involves lifestyle choices, personal consumption, manufacturing processes, organizational practices in public and commercial institutions, and community-based systems and programs. As Mazmanian and Kraft emphasize, "The problems of environmental pollution are now understood as inextricably tied to the way in which we produce and consume, and envision life's quality, from the individual up through the societal level."[44] In general, the sustainability movement has dodged many of the criticisms levied against environmentalism; however, there are some vocal conservative critics.

The sustainability movement in the United State has been shaped by the United Nations' sustainable development goals, first described in the 1992 Rio Declaration on Environment and Development known as Agenda 21. Referring to the "Agenda 21 Conspiracy," far-right private property rights groups such as the American Policy Center have drummed up anti-sustainability rhetoric and fought some local sustainability initiatives. Opponents view Agenda 21's principles of environmental sustainability as a threat to "lifestyles and consumption patterns of the affluent middle class—involving high meat intake, use of fossil fuels, appliances, home and work air conditioning, and suburban housing."[45] According to research conducted by the Southern Poverty Law Center, the Republican National Committee denounced Agenda 21 in their party platform in 2012.[46] The following year, Senator Ted Cruz claimed that Agenda 21 would abolish "golf courses and paved roads."[47] During her U.S. Senate campaign, Joni Ernst claimed that Agenda 21 would lead to the forced eviction of Iowa farmers, removing them from their privately owned agricultural lands to major urban centers.[48] The National Association of Scholars released a report in 2015 denouncing sustainability in higher education called "Sustainability: Higher Education's New Fundamentalism." They argue, "To the unsuspecting, sustainability is just a new name for environmentalism. But the word really marks out a new and larger ideological territory in which curtailing economic, political, and intellectual liberty is the price that must be paid now to ensure the welfare of future generations."[49] More recently, some red states have taken steps to actively bar the implementation of sustainability initiatives in state government by banning the adoption of "Environmental, Social, Governance" sustainability goals in state finance and investing.

Most fundamentally, conservatives' are protecting unfettered freedom in traditional ways of life in both the public and private sectors normalized by the DSP. According to Mark Stoddart et al.,

> Much of the anti-environmentalism that pervades the American right draws from a shared "deep story" that profoundly distrusts government intervention, trusts in the free market as emblematic of the American Dream, and asserts a defense of Christian faith, family, whiteness, and traditional masculinity against the political and cultural shifts provoked by the social movements of the 1960s protest cycle.[50]

Thus, in answer to what exactly are conservatives defending, conservatives are protecting a comprehensive worldview that sanctions a way of life unconstrained by environmental limits and progressive values. Environmentalism—the public face of the NEP—is a threat to a traditional way of life rooted in white, male, human supremacy over a complacent earth.

THE DRIVERS OF A PARADIGM SHIFT

The NEP Scale, the survey instrument that revealed the emergence of the NEP in American society, is a measure of public opinion. However, the ecological worldview embedded in the NEP is congruent with envi-

ronmental science. Thus, in environmental politics, the threat posed by
the NEP to defenders of the DSP is complex. The momentum of the environmental movement is powered by popular support and environmental
science. After the first Earth Day, the environmental movement emerged
as a potent force in society. According to Hazel Erskine, "Alarm about the
environment sprang from nowhere to major proportions in a few short
years."[51] The emergence of the environmental movement coincided with
growth in the environmental sciences; the environmental movement is
unique among social movements for its reliance on science.[52] Ecologists
had been studying the workings of ecosystems for decades; in the 1980s,
a new breed of biologists focused on biodiversity helped launch the field
of conservation biology. Similarly, scientists had been studying the global
climate long before the 1970s; the field of climate science took off in the
1980s. The growing body of scientific evidence corroborated public perceptions of environmental decline. This body of scientific evidence told
a story of environmental damage that directly challenges the view of the
world embedded in the dominant worldview. The ever-growing body of
science poses a threat as a driver of a paradigm shift in society.

Building on Thomas Kuhn's classic work on scientific paradigms,[53]
Hanna Cortner and Margaret Moote explain that a paradigm shift occurs
only when "a significant body of knowledge or information accumulates
that is contradictory to, or unexplained by, the accepted paradigm."[54]
Within scientific disciplines, paradigm shifts sometimes occur, although
they may be protracted and contested. For example, geologists once
thought the continents were immovable. Plate tectonics represented a
new scientific paradigm: the continents have moved. Eons ago, Africa and
South America were connected.[55] In the context of environmental politics,
environmental scientists have produced a massive body of evidence-based
consensus science that directly conflicts with assumptions about nature
and the resilience of the earth embedded in the DSP. Unbridled resource
exploitation and environmental despoliation cannot continue without
devastating ecological consequences.

One might think the avalanche of scientific evidence that thunders a
warning to humankind would have propelled a paradigm shift to the NEP.
However, dislodging a social paradigm is a more multifaceted endeavor
than a scientific paradigm shift. The dominant paradigm is sustained by
the power of society elites who can marshal multiple forces to preserve
and reinforce the dominant worldview. Society elites control major insti-

tutions and resources and have a powerful influence on the contours of public discourse. Cotgrove observed, "The struggle to universalize a paradigm is part of the struggle for power."[56] In other words, the struggle to displace the DSP is deeply political. As the public face of the NEP, the environmental movement represents a potentially powerful engine of political mobilization and social change.

The environmental movement also is a voice of science. The environmental movement and the science that undergirds the movement pose a direct challenge to the political and economic power of conservative elites invested in the DSP. Like the global warming-fueled storm surge flooding the streets of coastal cities such as New York City or New Orleans, the overwhelming body of consensus science produced by legions of environmental scientists in the past five decades represents an existential threat to conservative defenders of the DSP. To protect the DSP—a worldview that has brought us to the brink of planetary catastrophe—environmental science and the "voices" of science must be quashed. American conservatives have been attacking the legitimacy of the environmental movement, environmental science, and scientists since the 1980s to block a shift to a worldview grounded in ecological sanity.

Conventional political accounts do not tell this deeper story of environmental politics: a tale of competing paradigms and their defenders. Environmental advocates recognize the looming environmental catastrophe driven by the DSP worldview. A bipartisan environmental movement supported by a large—and potentially growing—majority of the American public paired with an overwhelming body of environmental science that contradicts assumptions about humans and the earth embedded in the DSP could drive a societal paradigm shift. From the perspective of the rearguard defenders of the DSP, the legitimacy of the environmental movement and the authority of environmental science have to be discredited in the eyes of public. Thus, we need to expand our analyses of environmental politics beyond partisanship, ideology, and the other explanatory variables of political analysis to incorporate the centrality of the earth and the subterranean role of the DSP and the actors defending it in our environmental politics. We can weave a richer story of environmental politics by connecting the well-orchestrated campaign to discredit the environmental movement, pervert environmental science, and smear the voices of science with the larger systemic dynamics of the ongoing paradigm shift driving society and environmental politics.

In recent years, some analyses of environmental problems and politics have expanded beyond traditional political variables and turned to critiques of free-market capitalism.[57] There is much truth in the argument that capitalism, with its relentless and delusional faith in infinite economic growth on a finite planet, is driving us into a world of irreversible ecological losses and planetary destabilization. Free-market capitalism is the cornerstone of the economic liberty cherished by conservatives. However, capitalism is an economic system embedded within the DSP; it is not a comprehensive worldview. Ultimately, it is the dominant worldview that prioritizes economic growth over environmental protection. It is the dominant worldview that prioritizes the relentless exploitation of natural resources and endless, unrestrained economic growth based on mythological claims of human exemptionalism and planetary limits.

How is it that we have missed or ignored the story of competing paradigms in environmental politics? This story is overlooked due to the influence of academic training and political habits. Shaped by the dominant assumption that humans are separate from nature, our politics and social science scholarship are largely divorced from nature. Social scientists focus on people. Most analyses of environmental politics typically focus on traditional political variables and issues including conservatives' opposition to environmental regulation and taxes, politicians' electoral fortunes, stakeholders' interests and political positions, and institutional and budgetary opportunities and constraints. We ignore the earth and environmental science.

Thus, we have not connected ideological opposition to environmental protection and regulation with the deeper story of competing worldviews and the purposeful distortion of the environmental movement and environmental science in our politics. Caricaturizing environmentalists with negative labels and unfounded accusations discredits the movement in the eyes of the public. Falsifying and contorting environmental science undercuts the case for environmental regulation and legislation. Politicizing environmental science undermines the authority of science and further weakens the environmental movement by discrediting the science that undergirds environmentalists' positions and the case for social and political change. Defenders of the DSP see environmental science and the story of ecological degradation it tells as an existential threat to the DSP worldview. Likewise, defenders of the DSP see the environmental movement—the public face of the NEP—as an existential threat to the American way of life. In fact, it is the DSP that poses a genuine existential threat to all of humankind.

A vast body of empirical evidence indicates that the DSP view of the world is scientifically inaccurate and completely out of touch with our planetary condition. We need to acknowledge that continued adherence to the DSP is downright dangerous. Fifty years ago, Dennis Pirages and Paul Ehrlich warned,

> The persistence of any society is threatened when its DSP no longer offers valid guidance for survival. . . . Today industrial society is threatened . . . by the uncritical acceptance of an outmoded DSP that cannot be sustained in the environment of the future.[58]

It is time to confront the ecological delusions of the DSP and the actors fiercely defending rearguard business and politics as usual. It is time to tell the deeper story of environmental politics.

TWO

AN EVENTFUL ANTHROPOCENE

Centering the earth in environmental politics and scholarship suggests that we should understand the planet that we must protect. The proposed geological epoch known as the Anthropocene is profoundly shaping modern life in the twenty-first century. Although discussions of the Anthropocene are becoming more frequent in academic circles, the reality of the proposed epoch is poorly understood. Common assumptions about the Anthropocene grounded in the dominant societal worldview threaten human and planetary well-being. Though the roots of the term suggest that the Age of Humans has arrived, the reality of the proposed epoch is better understood as the Planetary Era, a time when the Earth System has become an increasingly unruly actor in human affairs.[1] This chapter introduces the central importance of the Earth System in the Anthropocene conversation and the debate over the designation of the Anthropocene as an official new geological epoch. The chapter discusses the implications of the Anthropocene for the human experience on earth and illustrates how perceptions of the Age of Humans shaped by the dominant societal worldview are fallacious and dangerous.

THE EARTH SYSTEM

The Anthropocene demands a new understanding of the scale of human environmental impacts. As early as 1987, a report commissioned by the UN's World Commission on Environment and Development, titled *Our*

Common Future, foreshadowed human impacts at the planetary scale. The report cautioned that humans have the power "to radically alter planetary systems."[2] The phenomenon of change at the planetary level driven by human activities represents a distinctive, ongoing period in human history. First proposed in 2000 by the late Nobel Prize–winning atmospheric chemist Paul Crutzen, the Anthropocene has been described as an "unintended experiment of humankind on its own life support system."[3] Human activities are driving the Earth System outside of the parameters that defined the Holocene, the most recent post-glacial geologic epoch representing the past 10,000 to 12,000 years.

A common misunderstanding is that the Anthropocene is simply a new term to describe human changes to the environment. For decades, we have heard an all-too-familiar story of environmental despoliation, with countless examples of devastation and contamination seen and unseen. Wetlands, fertile soil, and beloved landscapes paved over for development. Ancient mountains blown apart to extract coal. Plastics—the underbelly of fossil fuels—literally everywhere, littering roadsides, fluttering from trees out of reach, and peppering most of the world's formerly pristine beaches. A literal soup of synthetic chemicals contaminates ecosystems on land and in water, eventually ending up in the world's oceans, the last stop for much human damage. Chemicals in countless commercial products mimic the hormone estrogen and interfere with fetal development in all kinds of organisms, including humans; males in some aquatic species exposed to artificial estrogens have literally been turned into females. Land conversions driven by population growth and assumed economic imperatives destroy vital habitat and contribute to the global biodiversity crisis. We don't need to look far on social media to encounter heartbreaking scenes of environmental and biological loss. A lone orangutan limping down the new dirt road made by the loggers who destroyed its forest habitat. An orphaned gorilla seeking comfort in the arms of a game warden. Elephants refusing to leave the butchered bodies of their relatives slain for ivory.

However, human impacts on the physical environment go beyond local or regional resource exploitation, destruction of natural habitats, and even globally dispersed pollution. In 2011, environmental writer Bill McKibben wrote *Eaarth: Making a Life on a Tough New Planet*.[4] McKibben argued that due to human-caused increases in atmospheric CO_2 and other greenhouse gases (as well as acid precipitation and deterioration of the ozone layer), there is no longer any place on earth free from human influence. We have so fundamentally transformed nature that we have left our familiar planet behind. We now live on Eaarth, a profoundly different, human-dominated planet.

However, McKibben's Eaarth provides only a partial understanding of the Anthropocene. Human impacts on the earth transcend local, regional, and even global impacts on nature. To understand the Anthropocene, we need to wrap our heads around the idea of the earth as a single complex system. As Clive Hamilton emphasizes,

> It is of utmost importance to understand that the "Anthropocene" is not a term coined to describe the continued spread of human impacts on the landscape or further modification to ecosystems; it is instead a term describing a rupture in the functioning of the Earth System as a whole.[5]

The Earth System is composed of the atmosphere, hydrosphere (water), cryosphere (ice), biosphere (ecosystems containing living organisms), and lithosphere (the earth's crust) and their interacting physical, chemical, and biological processes. A layperson's understanding of the Earth System involves recognizing that the Earth System is greater than the sum of its parts. Just as the human body is greater than the sum of its parts—our bodies are composed of multiple, interacting subsystems such as the respiratory, digestive, muscular, and skeletal systems—the Earth System is greater than the sums of its parts. In other words, as Earth System scientists explain, "Earth operates as a single, complex, adaptive system, driven by the diverse interactions between energy, matter and organisms."[6]

We need to understand the concept of the Earth System to grasp what is at stake in our Anthropocene journey. However, as a society, we are woefully ignorant about the Earth System. Beyond our ignorance, understanding the idea of the earth as a single complex system can be a major conceptual leap for many. Accustomed to thinking of earth as the "third rock from the sun," we typically view the earth as an inert substrate occasionally rocked by geological forces whose outer surface supports living species, including ourselves. We are not used to contemplating earth as a living entity, earth as a single system. As Hamilton notes, "Grasping the idea of the Earth System . . . requires a kind of gestalt shift . . . a big 'Aha' moment . . . without it the Earth is [inaccurately] understood as the aggregation of ecosystems more or less modified by humans."[7] In the Anthropocene, we are changing the behavior of the Earth System.

THE ANTHROPOCENE: EPOCH OR EVENT?

In 2009, a multi-disciplinary working group was convened to determine whether the Anthropocene should be formalized as a new geological epoch, and if so, to determine when it began. Earth System scientists argued that changes in the overall Earth System did not occur until the middle of the twentieth century. Driven by the rampant growth of the human enterprise dubbed the Great Acceleration, beginning in 1950, we can see exponential growth in major socio-economic trends such as global population, real GDP, and primary energy use that have caused a dramatic planetary-scale response, including an exponential increase in atmospheric CO_2, ocean acidification, tropical forest loss, and terrestrial biosphere degradation among other changes. In a visually compelling set of graphs, the Great Acceleration illustrates how profound and unprecedented change in socio-economic trends and planetary responses began shooting upward after 1950.[8]

In 2019, after a decade of research and robust debate, the Anthropocene Working Group recommended that the official start of the Anthropocene should be 1952, coinciding with the advent of nuclear fallout and the Great Acceleration. However, throughout the debate, some members of the working group argued that the Anthropocene began after the Industrial Revolution, when the use of coal first fired the upward trajectory of greenhouse gas emissions. Others argued that confining the onset of the Anthropocene to the past seventy years ignores the reach of human impacts on the earth including the effects of agriculture, colonialism, and industrialization that stretch back thousands of years. As William Ruddiman et al. noted, "a focus on the most recent changes risks overlooking pervasive human transformations of Earth's surface for thousands of years, with profound effects on the atmosphere, climate, and biodiversity."[9]

The working group was charged with evaluating the evidence for the onset of the proposed new epoch. The formal designation of the Anthropocene as an official new geological epoch will be determined by geologists on the Subcommission on Quaternary Stratigraphy. The subcommission's official designation will mark the onset of the Anthropocene in the stratigraphic record—literally in the rocks. Geologists drive a golden spike into the rockface to demarcate a new geological epoch for future generations. In 2023, scientists proposed the bottom of Crawford Lake in Ontario, Canada, as the site for the "golden spike."[10] However, in March 2024, the subcommission rejected the narrowness of the proposal in both time and space given that humans' transformation of the earth is a global story going back centuries. Subcommission members argued that the Anthropocene should be considered an ongoing "event" rather than an epoch, noting that the notion of geological events is well established in the scientific literature.[11] According to Philip Gibbard et al., viewing the Anthropocene as an ongoing geological event "more closely reflects the reality of both historical and ongoing human-environment interactions."[12]

Although geologists have rejected the proposal to formalize the Anthropocene as a new epoch with the official golden spike, the realities and dangers of the Anthropocene Event are widely supported within the multi-disciplinary scientific community. As Erle Ellis notes, "The Anthropocene is not an epoch—but the age of humans is most definitely underway."[13] Thus, the use of the term in both scholarship and cultural discourse is unlikely to disappear. More importantly, whether we refer to it as an epoch or event, we are living with the reality and consequences of the Anthropocene. The understanding of the Anthropocene central to our lived experience on this planet is based on the trajectory of the Earth System driven largely by the Great Acceleration. In 2007, Earth System scientists warned, "Human activities . . . are pushing the Earth into planetary terra incognita. The Earth is rapidly moving into a less biologically diverse, less forested, much warmer, and probably wetter and stormier state."[14] The Anthropocene Event should be understood as a protracted journey during which we are leaving behind our familiar planet.

What does it mean to enter "planetary terra incognita"? The significance and stakes associated with our journey away from our familiar planet become clear when we consider the current geologic epoch, the Holocene. All of modern civilization arose during the Holocene, the past 10,000 to 12,000 years. The unusually stable climate of the Holocene gave rise to modern agriculture, major cities, and socially and technologically complex societies. Although humans were living on the earth for thousands of years before the Holocene, everything that we treasure in modern civilization—agriculture, infrastructure, dynamic urban metropolises, complex technologies—arose because of the stable climate of the Holocene. The large coastal cities of the world exist because of the stable sea levels of the Holocene. Our globalized economy owes its existence to Holocene stability. Reliable access to natural resources, well-functioning global supply chains, and the security of financial systems depend on a stable climate.

Our nation's infrastructure was built to withstand the predictable climate of the Holocene. The sobering reality of the Anthropocene Event is that human impacts now threaten the stability of major planetary systems upon which all of modern civilization depends.

Still early in our journey into planetary terra incognita, we already are witnessing billions of dollars in damages as bridges wash away in floods, roads buckle in the heat, runways melt, and electricity grids fail in temperature extremes. Investments in traditional infrastructure will be dwarfed by the financial costs of repair and climate change adaptation. We face the prospect of Wall Street literally drowning in the rising seas. In coastal cities worldwide, billions of dollars of real estate investments will be washed away in the oncoming tides. Unabated global warming will likely cause financial chaos. Swiss Re, one of the world's largest providers of insurance, has warned that unchecked global warming and the resulting climate chaos could "Cut World Economy by $23 Trillion in 2050."[15] In our journey through the Anthropocene Event, we are sailing into an era of planetary destabilization, an irreversible, chaotic planetary reality that we will not recognize. David Orr warns, "Virtually everything that we presently take for granted will change, much to our disadvantage."[16]

AGE OF HUMANS OR PLANETARY ERA?

Like gazing in a funhouse mirror, viewing the Age of Humans through the lens of the dominant societal worldview offers a distorted picture of human possibility and prospects. The dominant worldview is based on fallacious claims about human dominion and control of nature. According to proponents of the dominant worldview, the Anthropocene offers an exciting new technological frontier in human control of nature. Reflecting this sentiment, journalist and author Mark Lynas asserts, "Nature no longer runs the Earth. We do."[17] In a similar vein, Michael Shellenberger and Ted Nordhaus emphasize, "The issue is not whether humans should control Nature, for that is inevitable, but rather how humans should control natures—nonhuman and human."[18] In the realm of climate change policy, engineering-based controls of the global climate collectively known as geoengineering represent the epitome of this ill-founded confidence in human command of nature. Faith in human control of nature is undergirded by the common assertion that we have reached the end of nature in the Age of Humans. However, believing we have reached the end of nature and are in control of the human-impacted ecosystems of the Anthropocene are illusions grounded in a dysfunctional worldview and a superficial understanding of our Anthropocene journey. The Anthropocene Event is better understood as the Planetary Era, a time when the Earth System is an increasingly unpredictable player in human affairs.

The End of Nature?

In his 1989 book *The End of Nature*, Bill McKibben argued that humans have so profoundly altered the physical reality of nature that we have reached the "end of nature."[19] Human-caused global warming touches every ecosystem on earth including the warming and acidifying oceans, the ultimate sink of much human damage. Book titles like *Coming of Age at the End of Nature* and *Living Through the End of Nature* are increasingly common in the scholarly and popular literature.[20] The "end of nature" pronouncement typically refers to the end of nature free from human influence. Although the idea of pristine nature is dear to many, in fact, beliefs about untouched nature are historically inaccurate. Informed by historical, archeological, and paleo-ecological records, we now know that most of what we think of as natural landscapes free from human influence have, in fact, been shaped by humans. For example, national parks such as Yellowstone, the Grand Canyon, and many others originally were home to native peoples and tribal nations who shaped their local environments in a variety of ways. Humans have been managing their surrounding landscapes for thousands of years. With practices including hunting, burning, introduction of exotic species, and cultivation, we have scrambled the species composition of ecosystems all over the world and transformed an estimated three-quarters of the earth's landmass.

The "end of nature" declaration also is based on a limited understanding of nature. We are accustomed to thinking about nature as a tangible, geographic place we can observe and experience. To understand the workings of nature, we need to expand conceptualizations of nature to include ecological processes at all scales, including the behavior of the Earth System. In other words, we need to understand nature as a verb, *nature as process*. Recall that the Earth System is composed of interacting physical, chemical, and biological processes among the atmosphere, hydrosphere (water), cryosphere (ice), biosphere, and lithosphere (rocks). Although human activity is now a significant force influencing the trajectory of the Earth

System, as Mark Maslin and Simon Lewis emphasize, "the fundamental processes governing the Earth System are the same now as in the past."[21]

Thinking about nature as process is more intangible and understandably difficult for the average person to comprehend. It involves distinguishing between human impacts and the responses of natural systems. For example, global warming is causing extensive die-offs in most of the world's warmwater coral reefs. By some estimates, coral reefs could disappear by mid-century. As a symbiotic organism, coral is a combination of a coral polyp and algae. The coral polyp offers structural support for the photosynthesizing algae that feed the coral and give corals their dramatic colors. Human impacts are stressing coral, causing the coral to expel the algae in a process known as coral bleaching. Once the algae are gone, the exposed coral substrate appears ghostly white, hence bleaching. At the simplest level, we could argue that humans are causing coral bleaching and die-back. A more nuanced understanding differentiates between impact and response. Humans are causing the impacts, but we do not control the corals' response. In other words, to understand nature as process, we need to understand ecological processes and functions that continue to operate even in the human-impacted ecosystems of the Anthropocene Event.

Nature in Novel Ecosystems

One of the hallmarks of the Anthropocene is the widespread distribution of human-impacted ecosystems known as novel ecosystems. Some might conclude that novel ecosystems are the physical embodiment of the end of nature. Novel ecosystems contain new species combinations and abundance that did not occur in previous, historic landscapes; they are the result—often unintended—of human disturbances.[22] A wide range of human disruptions lead to the development of novel ecosystems including land use changes such as modern industrial agriculture, dam building, timber harvesting, quarrying, and soil removal as well as intentional and unintentional chemical inputs (such as nitrogen deposition) that change soil composition.[23] Novel ecosystems often displace native ecosystems like forests and grasslands that may have important ecological or nostalgic value to people. In the context of conservation biology and restoration ecology, novel ecosystems are problematic and controversial. They pose challenges for the conservation of traditionally valued aesthetic and ecological properties in historic landscapes. However, there is an important difference between what humans value about traditional ecosystems and landscapes, many of which are being displaced by novel ecosystems, and how novel ecosystems function.

When we focus on ecological processes such as nutrient cycling and biomass production, novel ecosystems provide important ecosystem services. They also share key characteristics of wild ecosystems including constant change, uninhibited growth, and evolutionary processes.[24] As Nathaniel Morse et al., emphasize, "All ecosystems, including novel ones, change because of natural processes."[25] In addition, non-native species found in novel ecosystems can provide habitat and food sources for other species; novel ecosystems also can support the regrowth of native species.[26] Novel ecosystems contribute to the functioning of the overall Earth System in a variety of ways including photosynthesis, nutrient cycling, and soil formation. The behavior of novel ecosystems—*nature as a verb*—alerts us to an important reality that directly conflicts with common assumptions about the end of nature in the Age of Humans. The behavior of both historic and novel ecosystems shows us that nature has agency.

In reality, nature always has had agency and the ability to profoundly disrupt human affairs. However, the stability of the Holocene paired with assumptions embedded in the DSP have seduced people into believing that the earth is a passive entity that we can control with technology in the Age of Humans. Moreover, the comforts and convenience of modern civilization usually insulate us from the forces of nature. However, our journey away from the stable conditions of the Holocene—our journey into the global warming-fueled storms, wildfires, and floods of terra incognito—demands a new understanding of nature's agency. Nature is an increasingly active and dangerous force to be reckoned with in the Planetary Era.

Human Power Versus Control

In our journey out of the Holocene, we have not arrived at the end of nature. Quite the contrary. The stability of the Holocene obscured the reality of nature's agency. In the Anthropocene, the Earth System is an increasingly active, unpredictable player in human affairs, a reality that challenges deeply held assumptions about human power and control embedded in society's dominant worldview. To understand the reality and dangers of the Anthropocene, we need to comprehend that despite our immense powers, we do not have control. We can change, disrupt, and interfere with nature, but we cannot control nature. As Anne Fremaux warns, we must accept "the non-mastery of our technological mastery . . . humans are not able to control their control of nature, nor can they control their own lack of control."[27] Visions of planetary control are a fantasy, a dangerous illusion powered by a worldview entranced by technology and deeply disconnected from ecological realities.

In the Planetary Era, nature is not the passive, manageable backdrop for human affairs imagined by the dominant worldview. As Fremaux points out, nature is

> a self-causing reality, an active process, and a self-productive power that ceaselessly generates new forms. . . . At the very time that nature is declared "dead," natural processes are making their comeback in the form of uncontrolled phenomena.[28]

We experience nature as an active, unpredictable force in countless ways. Blinded by technological optimism, we repeat the same patterns over and over. Ignoring Rachel Carson's warnings decades ago, we spray almost a billion pounds of pesticides on agricultural pests every year only to watch the pests evolve and become resistant to the very same pesticides we trusted. So, we create more pesticides, only to see the cycle repeated. We douse livestock in giant feedlots with antibiotics only to discover bacteria are becoming resistant to our lifesaving medicines. Viruses like COVID-19 mutate and wreak havoc on the global population. We think we are in control, but we live in a world of unforeseen consequences. Nature is active and unpredictable.

Many have assumed global warming is like a thermostat with our hand on the dial. We thought we were gradually and predictably warming the climate, slowly enough to detect problems and fine-tune technologies that would enable us to continue to burn fossil fuels in defiance of the laws of physics. We thought global warming would simply mean a warmer planet as we gradually cranked up the thermostat. Global warming sounded benign, comfortable. Instead, global warming has unleashed countless unforeseen consequences. Nature is active and unpredictable.

Who would have guessed that the rapidly warming Arctic and altered atmospheric circulation patterns might cause wavy southward dips of polar air—the infamous polar vortex—causing frigid temperatures, social and economic disruption, suffering and even death in unlikely places, like we saw in Texas during the winter of 2021? Did we anticipate that global warming would cause climate chaos, fiercely destructive storms in all seasons, including bomb cyclones of snow and record-breaking monsoon rains? Those familiar with the fundamentals of photosynthesis thought global warming would simply mean a greener, more productive planet. We thought higher levels of CO_2 would boost plant growth, never suspecting that dramatically increasing atmospheric CO_2 could make staple crops like wheat and rice less nutritious. Did we ever imagine that ancient viruses entombed within glaciers, ice sheets, and permafrost could be released in a warming, melting world, posing novel threats to human health? Nature is active and unpredictable.

Human interventions in the climate system, known as geoengineering, conjure up visions of even greater control of the global thermostat. We assume we can predictably cool the earth by turning the dial in the opposite direction. However, while it is possible to cool the global climate with geoengineering, we cannot control the results. Some regions of the earth could warm even faster than they would have warmed without geoengineering with potentially dangerous consequences. If the poles warmed even more because of geoengineering, it could further destabilize West Antarctica or the Greenland ice sheet and speed up sea-level rise. In other words, if we turn down the global thermostat using geoengineering technologies, we have no way of knowing whether the results will improve or worsen our planetary dilemma.

It turns out that the more appropriate metaphor for global warming and climate change is a switch. But our hand is not controlling the switch; change can be abrupt and surprising. Natural systems can shift abruptly once the system crosses a tipping point. A tipping point is defined as a threshold where "small changes become significant enough to cause a larger, more critical change that can be abrupt, irreversible, and lead to cascading effects."[29] Climate scientists have identified numerous tipping points in the global climate system; we are moving dangerously close to several of them. In 2021, almost 3,000 scientists issued a declaration of planetary emergency: "There is also mounting evidence that we are nearing or have already crossed tipping points associated with critical parts of the Earth system, including the West Antarctic and Greenland ice sheets, warm-water coral reefs, and the Amazon rainforest."[30] We are used to thinking about global change in geological terms occurring gradually, over the course of millennia. As the term "tipping points" suggests, changes in the global climate system can occur quickly, in years rather than centuries.

In 2013, the National Academy of Sciences released a report titled

"Abrupt Impacts of Climate Change," which explained that "The history of climate on the planet—as read in archives such as tree rings, ocean sediments, and ice cores—is punctuated with large changes that occurred rapidly, over the course of decades to as little as a few years."[31] Wallace Broecker's research on ocean circulation and climate change led to the stunning discovery that thousands of years ago, the North Atlantic "conveyor belt" that transports heat from the Gulf Stream, warming northern regions of the globe, shut down abruptly, plunging northern Europe into frigid conditions within just a few years.[32] *Within just a few years.* In 2013, the National Academy of Sciences warned that sudden change in "ecosystems, weather and climate extremes, and groundwater supplies critical for agriculture now seem more likely, severe, and imminent."[33] Abrupt climate change has major implications for climate change adaptation and human welfare. The National Academy of Sciences report also cautioned that the global climate can shift so suddenly that by the time we recognize the problem, adaption would be virtually impossible, leading to social and economic chaos.

Change in complex systems like the global climate system is unpredictable, driven by multiple forces and feedbacks that we do not fully understand or control. Technology-based environmental controls such as geoengineering are grounded in mechanistic thinking that assumes we can command the global thermostat and predict and control the outcomes. However, geoengineering is poorly adapted to the world of complex systems, a world of tipping points, sudden change, and surprise—a world in which nature is an active and unpredictable force. As climate scientist Michael Mann emphasized, "The fundamental problem of geoengineering solutions is the monumental danger of tinkering with a complex system that we don't fully understand."[34]

Global warming with all its surprising consequences was an unintended experiment. Why would we assume we could geoengineer our way to a stable climate without unleashing even more surprising and damaging effects? The repercussions of geoengineering will be shaped by the behavior of natural systems. As ecologist Camille Parmesan cautions, "Once some of these processes get chugging along, they may reach a point where it becomes impossible to stop them. Humans can control human actions, but humans cannot control the biosphere's responses."[35] In the Planetary Era we are rushing headlong into a world of "unknown unknowns."[36] Our future will be forged by how natural systems act and react to human disturbances. In spite of the unshakable confidence and hubris driving much thinking about geoengineering technologies, we will never know enough about the workings of nature to control the global climate system.

If humanity is to thrive in the Planetary Era, we must confront the perilous delusions spawned by adherence to a worldview grounded in the certainty that we can control nature. As Diane Dumanoski observed, "For four centuries, it did not seem to matter that the vast construct of modern civilization rested on an inaccurate view of nature. . . . Now it matters most of all."[37] It is time to jettison a worldview grounded in mythological visions of technological control of an amenable earth. Believing we are in control of nature is a fantasy. Nature is not a compliant partner in the pursuit of human aspirations. Nature is an active, unpredictable force. As Fremaux cautioned, nature is not "a pure raw material that awaits human inventiveness and ingenuity passively. It is a complex system that can wake up at any time."[38] Wallace Broecker warned, "the climate system is like an angry beast and we are poking it with sticks."[39] In the Planetary Era, the Earth System is wide awake.

IRREVERSIBILITY IN THE ANTHROPOCENE

In our journey through the Anthropocene Event, we already have incurred substantial human, economic, social, and ecological losses. However, in politics and education, we never address the implications of a distinctive type of harm caused by economic activities and government policies: irreversible ecological loss. In the Anthropocene Event, we are faced with irreversible ecological losses such as species extinction, unstoppable melting of glaciers and polar ice caps, and vanishing coastlines. Irreversibility compromises future generations' ecological inheritance and contravenes democratic norms.

Seventeen hundred of the world's top scientists issued a stark warning in 1992: "Human beings and the natural world are on a collision course. Human activities inflict harsh and often irreversible damage on the environment and on critical resources."[40] Yet, in our politics and in education, we have paid scant attention to scientists' warning about irreversible damage, a unique feature of environmental policy of paramount importance. By ignoring irreversibility, we are not telling the full story of environmental policy and politics. We are not adequately preparing our students and the public for the planetary realities of our Anthropocene journey.

Policy decisions sometimes cause irreversible consequences. For example, in Flint, Michigan, many children's lives have been forever compromised by lead poisoning; they will live and struggle with intellectual disabilities and behavioral disorders for their entire lives. However, government policy decisions and actions by private commercial interests often result in irreversible ecological impacts and losses for all future generations on vastly different temporal scales. For example, as we worked our way across the North American continent, on both public and private lands from coast to coast, we destroyed entire ancient forest ecosystems. Old-growth forest ecosystems have unique structures and features that enable them to play crucial roles in wildlife habitat and biodiversity protection, hydrological regimes, nutrient cycling, and carbon storage—to name a few essential ecological functions. The vast majority of old-growth forests are gone for generations, destroyed forever on any meaningful human time scale. In the Pacific Northwest where abundant moisture enables trees to grow to towering sizes, we have destroyed millions of acres of ancient trees, jeopardizing countless species dependent upon the old-growth ecosystems, disrupting habitat for salmon and other prized sports species, and destroying the forests' ability to secure soil and absorb and regulate the release of the huge amounts of rain characteristic of the region.

In the southeast, intact old-growth forest ecosystems formerly provided flood control, nursery habitat for commercial fisheries, water quality protection, and habitat for many unique species. The razing of the ancient forests in the south caused the extinction of the Ivory-billed Woodpecker whose sonorous drumming formerly echoed throughout the old bottom-land forests of the south. Called the "Lord God Bird" by locals, the majestic Ivory-billed Woodpecker had a thirty-inch wingspan and magnificent red crest. Future generations of birdwatchers will never again be transfixed by the Lord God Bird. Although logging has destroyed most of the old-growth forest ecosystems in the Appalachian Mountains, they still house one of the highest concentrations of biodiversity in the United States. Mountain-top removal mining in Kentucky and West Virginia destroys biodiversity and obliterates geological landforms for all future generations, converting the ancient mountain peaks into barren, flat-topped plateaus crisscrossed with roads.

Even something as mundane as urban sprawl causes irreversible losses. As suburbs surrounding midwestern cities such as Chicago, Columbus, and Indianapolis have metastasized, gobbling up millions of acres of productive farmland, billions of tons of the best topsoil on the planet have disappeared, gone forever. Conventional farming practices also exacerbate irreversible losses. By some accounts, one-third of the original topsoil of the fertile midwest has disappeared, washed into the Gulf of Mexico, gone forever on any meaningful human time scale. The proliferation of toxic chemicals dispersed throughout the environment leads to different kinds of irreversible losses and impacts. So-called forever chemicals (per- and poly-fluoroalkyl substances referred to as PFAS) cause immutable and dangerous toxic pollution. Forever chemicals consist of chains of some of the strongest, most stable bonds known to chemistry; they are common in countless substances such as Teflon cookware. The ecological impact of forever chemicals needs to be understood on geologic time scales; they take thousands of years to degrade in the environment—thousands of years to degrade while permanently compromising the lives of countless people and species of wildlife.

And then there is global warming. Climate change poses irreversible, existential threats to ecosystems and human civilization. According to the IPCC's Sixth Assessment Report, "Many changes due to past and future greenhouse gas emissions are irreversible for centuries to millennia, especially changes in the ocean, ice sheets and global sea level."[41] Even if global warming ceased today, Greenland would continue to lose ice for millennia after greenhouse gases are stabilized. Globally, we are losing roughly 1.2 trillion tons of ice annually from Greenland, the Arctic, Antarctica, and mountain glaciers. And the seas are continuing to rise.

Irreversibility invokes temporal scales that pose vexing political problems that violate longstanding democratic norms. Policy reversibility is the norm in American politics; it is essential to the pact we make with future generations. Mathew Humphrey explains that the ability to revisit policy choices is central to the legitimation of democratic decision-making. He writes, "the democratic conversation does not cease with any one policy choice . . . the open democratic system ensures that any policy area can be revisited."[42] In other areas of domestic policy such as taxation or social welfare policy, policies can be revisited in subsequent congressional sessions or presidential administrations; the "democratic conversation" does not end. Irreversibility means that we get no second chances; the "democratic conversation" comes to a screeching halt. Ignoring irreversibility is one of the most serious failures in our politics. Students and members of the public—whose own lives already are being touched by irreversible environmental losses—need to understand the real stakes in environmental policy debates. In the real-world context of environmental politics, exploring the implications of irreversibility needs to be a central priority in our education and politics.

With few exceptions, analysts ignore irreversibility in environmental politics. Mathew Humphrey and Steve Vanderheiden discuss policy irreversibility in the context of radical environmental politics.[43] Both argue that irreversibility justifies the use of direct-action tactics as legitimate forms of political resistance. In the face of irreversible losses, activists often have only one chance to protect important resources and ecosystems. For example, individuals working alone and sometimes under the loose umbrella of Earth First! adopted controversial tactics known as ecological sabotage or "ecotage" to prevent the destruction of the ancient forests of the Pacific Northwest in the 1980s and 1990s. Earth Firsters! drove spikes into trees marked for harvest, then prominently advertised in spray paint, "These trees are spiked!" The activists were trying to prevent the irreversible destruction of the centuries-old giants that anchor old-growth forest ecosystems. They counted on the timber companies to leave the trees standing to protect mill workers who could be grievously injured by flying metal in a timber mill.

Numerous times, various acts of ecotage bought the time necessary for more mainstream environmental activists to seek administrative or judicial relief to stop illegal timber harvests. Many of the formal appeals and lawsuits filed by regional, mainstream environmental groups blocked Forest Service decisions described by one court as a "deliberate and systematic refusal . . . to comply with the laws protecting wildlife."[44] In other areas of policy, when illegal agency decisions are challenged in court, usually it is possible to correct or reverse policy mistakes. When it comes to irreversible ecological losses—like cutting down ancient forests—there are no second chances. Yet we never hear about the connections between radical environmental activism and irreversibility in the media. Ignorant of the high stakes associated with the resources defended by these activists, we label them "eco-terrorists."

Environmental sustainability is one of the three core dimensions of sustainability. Stephen Dovers argues that environmental sustainability problems are significantly different from other policy problems, with irreversibility being one of their distinctive features. He highlights the crucial fact that at the point of policy development and adoption, we often do not know the value of the ecological resources that are irreversibly destroyed.[45] In many areas of public policy, we cannot accurately predict the consequences of our decisions. Policy irreversibility significantly raises the stakes of our ignorance.

There also is some discussion of irreversibility in the growing literature on the Anthropocene. Several authors have touched on the governance challenges posed by irreversibility and the implications for intergenerational equity. Echoing Humphrey's concerns about policy reversibility in democratic decision-making, Hamilton et al. noted, "We have never thought about how to govern the irreversible."[46] Hamilton et al. also discuss the connection between irreversibility and intergenerational equity in the Anthropocene: "With the Anthropocene, this kind of undoing is no longer possible. . . . Many future generations have been thrown into the new era."[47] Ewa Bińczyk writes about the enormously high stakes associated with political and economic decision-making in the Anthropocene in the context of irreversibility. She points out that irreversible losses ultimately constrain the range of available options for meeting present and future human needs, putting the issue of intergenerational equity in sharp relief.[48]

In our politics, we pay scant attention to intergenerational equity and ignore the real-world implications of irreversible environmental policy decisions for future generations. Our politics are notoriously shortsighted, tied to election cycles and funding, and shaped by humans' general difficulty in focusing on long-term futures. We live with the consequences of our shortsightedness when we fail to make needed investments in education, health care, or infrastructure that can enhance quality of life for living generations. Future generations will be forced to live with the consequences of our shortsightedness if we continue to ignore the irreversible effects of our policy decisions. As Hamilton et al. suggest, irreversibility "should force us to rethink government."[49]

Ironically, looking backwards, Thomas Jefferson was concerned about intergenerational equity and the earth. In a two-year correspondence with James Madison discussing the debate over the future Bill of Rights, Jefferson wrote:

> I set out on this ground . . ."that the earth belongs in usufruct[50] to the living": that the dead have neither powers nor rights over it. . . . Then no man can, by natural right, oblige the lands he occupied, or the per-

sons who succeed him in that occupation, to the payment of debts con-
tracted by him. For if he could, he might, during his own life, eat up
the usufruct of the lands for several generations to come, and then the
lands would belong to the dead, and not to the living.[51]

Our politics and economics have been singularly focused on a neoliberal
interpretation of Jefferson's first phrase: *"the earth belongs . . . to the living."*
To the detriment of all future human generations, species, ecosystems,
and even the functioning of planetary systems, we have disregarded the
actual meaning of Jefferson's crucial insight. By discounting the future, we
have already consumed or destroyed "the lands for several generations to
come," compromising future generations' opportunities and welfare. We
are long overdue in confronting irreversibility and its ethical implications
in democratic politics.

Our journey through the Anthropocene is taking us into dangerous
territory. We are hurtling toward a future marked by planetary instability. Current generations already are living with the increasingly serious
consequences of altering the Earth System. However, the stark realities
of the Anthropocene launch questions of intergenerational equity to unparalleled heights. Our children have no hope of building a meaningful
future in a society driven by misguided fantasies about our ability to control nature. Our children face an uncertain future on a volatile and turbu-
lent planet. Based on current trajectories, we will likely experience social
and economic disruption, mass migration, and geopolitical upheaval. Add
potential food and water shortages, and the future looks grim. If we choose
to ignore the realities of the Planetary Era, our children will inherit an im-
poverished world of irreversible losses and destabilized planetary systems.
If we choose to ignore intergenerational equity in environmental politics,
the American dream will belong to the wealthy and the dead.

THREE

THE ASSAULT ON ENVIRONMENTALISM

Environmental politics is not simply a story of clashing political ideologies and policy preferences. Environmental politics is a story of clashing worldviews, each with fundamentally different assumptions about humans and nature and the gravity of our environmental problems. In this clash of worldviews driving environmental politics, the environmental movement plays a central role. The environmental movement represents the public face of the new ecological paradigm. The combination of a popular bipartisan social movement informed and motivated by a growing body of environmental science poses a twin threat for conservative defenders of the dominant worldview. For decades, conservatives in government, media, think tanks, and evangelical groups have been engaged in a two-pronged assault on the environmental movement. Arguing that environmental problems have been wildly exaggerated by environmental extremists, conservatives have politicized the movement by characterizing it as a radical left-wing political movement; they have discredited the movement by denigrating environmentalists with a slew of slurs and derogatory caricatures. The other path of assault involves denying, distorting, and politicizing environmental science. Undermining public confidence in environmental science further weakens the environmental movement. This chapter tells the story of conservatives' efforts to delegitimize the environmental movement, and it examines the implications of discrediting the environmental movement, an important voice of science and engine of potential social and political change.

INTRODUCTION TO THE ENVIRONMENTAL MOVEMENT

Our oldest environmental group, the Sierra Club, was founded in 1892 by famed naturalist John Muir and a small group of mountaineers interested in exploring the Sierra Nevada mountains. Beginning in the early 1900s, the Sierra Club worked with government officials in the creation of the U.S. Forest Service and the U.S. Park Service. Two Boston women, outraged at the killing of birds so their plumage could adorn women's hats, were instrumental in the founding of the Audubon Society in 1905. The Wilderness Society and the National Wildlife Federation joined the ranks of conservation organizations in the mid-1930s. These groups and others were primarily concerned with preserving and protecting nature and wildlife. Led by amateurs, these groups relied on the intrinsic beauty and characteristics of public lands and wildlife to garner public support. The Sierra Club became well-known for its Outings Program, designed to expose members to the grandeur of the Sierra Nevada range. The Outings Program brought new members into the Club who were interested in their hiking program and facilitated the expansion of the Sierra Club across the nation as a grassroots network of state and local chapters.

The Outings Program supported the Club's reliance on a conservation strategy known as "bearing witness."[1] Bringing people into the wilderness went beyond recreation. Deep immersion in nature—"bearing witness" to the beauty and unique features of America's natural heritage—was intended to motivate people to fight for special places they loved. In the 1950s, the Bureau of Reclamation developed a proposal to flood parts of the Colorado River within Dinosaur National Monument. In order to protect Dinosaur National Monument, Sierra Club leaders agreed to a compromise that enabled the Bureau of Reclamation to flood Glen Canyon to create Lake Powell. Unfortunately, Sierra Club members had not borne witness to the canyon's magnificent red rocks and geological formations. When they realized they had traded away one of our national treasures, club leader David Brower described Glen Canyon as *The place no one knew.*[2] Spurred on by this heartbreaking loss, the Sierra Club ramped up efforts to expand the reach of "bearing witness" as a conservation strategy to protect Dinosaur National Monument. The club encouraged visitation, published the book *This Is Dinosaur,* and released a film about Dinosaur.[3] In 1950, about 13,000 people visited Dinosaur National Monument. In 1954, after the Sierra Club's advertizing blitz, 51,000 people visited Dinosaur.[4] With pressure from the Sierra Club and other conservation groups, Dinosaur was saved by congressional legislation passed in 1956 that banned construction of dams or reservoirs in all national parks and monuments.[5]

Over time, the environmental movement needed to adapt to changing environmental and political circumstances and opportunities. The first generation of conservation issues required little scientific expertise. One only had to look at the Stegosaurus fossils embedded in the rocks at Dinosaur or gaze at the spectacular colors and vistas of the Grand Canyon to want to advocate for their protection. However, the birth of the modern environmental movement in the late 1960s ushered in a new generation of problems. These second-generation air and water pollution issues required considerable scientific expertise and could not be protected through a one-time act of Congress. Based on scientific research, Rachel Carson's landmark book *Silent Spring* was a harbinger of a new environmental movement to come grounded in science.[6]

If environmental groups were to advocate for clean air and water, they needed scientific expertise in a variety of natural science and public health disciplines. Soon, environmental organizations sought out the expertise of economists, lawyers, and policy analysts. Founded in 1967, the Environmental Defense Fund (EDF) went to court to protect Bald Eagles, Ospreys, and other fish-eating raptors from the effects of DDT.[7] Founded by a small group of lawyers in 1970, the Natural Resources Defense Council (NRDC) developed its own internal team of environmental lawyers, scientists, and policy experts working to pass "evidence-based environmental laws."[8] The large environmental groups advocated for the passage of landmark laws like the Clean Air Act,[9] Clean Water Act,[10] Safe Drinking Water Act,[11] and Endangered Species Act,[12] which were signed into law in the early 1970s.

This new foundation of laws paired with the creation of the EPA led to the development of a federal environmental policy system. Adapting to this new political opportunity, the mainstream groups became permanent fixtures in Washington as professional advocates for environmental protection.[13] Many of these groups established regional and state offices as well. Still in existence and active in environmental politics to this day, mainstream groups like EDF, NRDC, the Sierra Club, the Wilderness Society, the National Wildlife Federation (NWF), the Audubon Society, and others employ large staffs of natural scientists, public health experts, economists, policy analysts, and lawyers, all devoted to protecting wild places and people from environmental degradation and pollution of all kinds. They also regularly consult with a constellation of environmental scientists and researchers in academia and non-profit organizations. The environmental movement's grounding in science is one of its distinctive features.

The environmental movement continues to evolve in response to changing environmental, political, and social demands. Just as the movement had to adjust to the challenges of the second-generation pollution issues that arose in the late 1960s, the movement has adapted their strategies and organizational agendas to address growing demands for environmental justice. When the environmental justice movement burst onto the political scene in the 1980s, the mainstream groups were stridently criticized for ignoring the problems of the urban poor and for failing to incorporate more people of color in their staffs and leadership structures. The headline of a 1990 article in the *New York Times* proclaimed, "Environmental Groups Told They Are Racists in Hiring."[14] Shaken by the accusation, the environmental mainstream has sought to diversify their staffs and boards, and to broaden their agendas.

For example, in response to the *New York Times* article, in 1993, the Board of Directors of the Sierra Club adopted an environmental justice policy, stating, "The Board calls on all parts of the Club to discuss and explore the linkages between environmental quality and social justice, and to promote dialogue, increased understanding and appropriate action."[15] The Sierra Club's principles of environmental justice reflect the blending of their traditional concerns with contemporary environmental realities. Hearkening back to their founder, they note that John Muir said, "Everybody needs beauty as well as bread, places to play in and pray in, where nature may heal and give strength to body and soul alike."[16] NWF created the Environmental Justice, Health and Community Revitalization Program;[17] NRDC established the Environment, Equity & Justice Center.[18] Although many commentators would agree that the environmental mainstream still has a ways to go in advancing environmental justice, it is making progress. The mainstream agenda now incorporates a host of urban environmental issues. In other words, the "environment" now encompasses everything from wilderness and natural ecosystems to the urban spaces where, in the blended words of John Muir and environmental justice scholar Robert Bullard, people "live, work, play, and pray."

In response to the growing attention to environmental justice, in recent years, the public narrative about climate change has shifted from simply mitigating greenhouse gases to advancing climate justice. For example, EDF's climate change advocacy explicitly addresses climate justice.[19] Under the umbrella of environmental justice, other groups like NRDC have climate adaptation policies focused on "boosting public health infrastructure, occupational protections for workers on the frontlines, and developing heat action plans."[20] The older environmental groups have been joined by a host of new organizations and coalitions focused on climate justice such as the Climate Justice Alliance, 350.org, the youth-led Sunrise Movement, and the Citizens Climate Lobby. These groups and others, including the mainstream, recognize that billions of the world's poorest people living in low-lying coastal and arid desert regions are the most vulnerable to the impacts of climate change. The injustice is compounded by the fact that these populations have contributed almost nothing—measured in greenhouse gas emissions—to the problems wreaking havoc with their lives.

THE CONSERVATIVE ASSAULT ON THE ENVIRONMENTAL MOVEMENT

Conservatives' war on environmentalism began soon after the birth of the modern environmental movement. After the publication of *Silent Spring* ignited the American imagination, two accidents at opposite ends of the country in 1969 further stoked Americans' worries and emerging environmental consciousness.[21] Horrified by the sight of millions of gallons of crude oil spewing into the Pacific Ocean despoiling the pristine Santa Barbara beaches, citizens launched into action. On the other side of the country, Lake Erie was pronounced "dead" and Cleveland's Cuyahoga River caught on fire (one in a series of fires on the river). Images of the burning river blazed across the country on the cover of *Time* magazine, stoking widespread public concern about water pollution. All across the nation, many rivers and streams supported flotillas of litter. Countless Americans saw the famous Keep America Beautiful public service announcement on TV featuring a Native American, with a tear running down his face, guiding his canoe through a litter-choked stream past a factory emitting a trail of smoke.

By the end of the decade, the United States had witnessed a dramatic surge in public support for environmental protection. The public symbol of this energetic new movement, Earth Day 1970, attracted an estimated 20 million Americans who came out in force in communities all across the country to rally for environmental protection. Hazel Erskine commented on how quickly the environmental movement emerged as a potent force in society, writing, "A miracle of public opinion has been the unprecedented speed and urgency with which ecological issues have burst into American consciousness. Alarm about the environment sprang from nowhere to major proportions in a few short years."[22] The extraordinary growth of public support for environmentalism in American society was paired with bipartisan political leadership.[23] The Nixon administration launched a new era in environmental protection proclaiming, "Shall we make peace with nature and begin to make reparations for the damage we have done to our land, to our air, and to our water?"[24] President Nixon created the U.S. Environmental Protection Agency in 1970 and signed an unprecedented number of landmark federal environmental statutes in the early 1970s.

Lessons from the Reagan Administration

Conservatives glimpsed the threat posed by the fledgling environmental movement almost immediately. Researchers studying public opinion on environmental issues in the 1970s found that significant numbers of Americans were embracing an alternative ecological worldview. The environmental movement became the public face of that new worldview. Environmentalists' concerns about environmental degradation and the severity of the looming ecological crisis aligned with the new ecological paradigm. The burgeoning body of scientific evidence on environmental degradation and human health threats undergirded assertions embedded in the alternative worldview voiced by environmental activists.

Conservatives began mobilizing economically, politically, and culturally against environmentalism soon after Earth Day 1970. Some commentators quickly blamed environmentalists for the economic crises of the 1970s.[25] On the heels of President Jimmy Carter's ambitious environmental agenda, conservative opposition to environmentalism gained momentum with the election of Ronald Reagan. In 1982, conservative Republicans in the House of Representatives released an internal report called "The Specter of Environmentalism" that highlighted conservatives' underlying concerns about environmentalism. The report described environmentalists as extremists "who posed a growing threat to the orderly development of the nation's resources." The report went on to emphasize that environmentalism was no longer simply about protecting the environment, adding that environmentalism consisted of "an entire outlook of broad political and social affairs."[26] James Watt, Reagan's secretary of the interior, publicly denigrated environmentalists, claiming "What I call 'commercial' environmentalists are hard-core, left-wing radicals, manipulating the press. . . . They have a conspiracy of shared values."[27]

Both Watt and the House Republicans targeted conservatives' fundamental problem with environmentalism: their references to "an entire outlook of broad political and social affairs" and "a conspiracy of shared values" hint at their recognition of an inherent, subterranean worldview involving profound social, cultural, and economic change associated with environmentalism.

Linking environmentalism with an alternative worldview has been repeated over the years by a variety of conservative opponents of environmentalism and continues to the present. For example, Calvin Beisner, the founder of the Cornwall Alliance—a conservative Christian evangelical, free-market group—refers to environmentalism as "a complete alternative world view," a "native evil" that constitutes a threat to western civilization.[28]

After decades of bipartisan cooperation on environmental policy, Reagan set a new course for the Republican Party. Driven by a conservative agenda rooted in the dominant worldview, the Reagan administration set out to weaken environmental protection through regulatory reform, budget cuts, and personnel policies that installed ideological loyalists.[29] Reagan's environmental policies resulted in the loss of thousands of experienced personnel and huge reductions of 50 percent or more in the research budgets of the EPA and other environmental and natural resource agencies.[30] Stories abound about the anti-environmental positions and missteps of Secretary of the Interior James Watt and the Administrator of the EPA Anne M. Gorsuch. Gorsuch was a willing partner in Reagan's regulatory reform initiatives and budget slashing at the EPA. Watt's positions on public land management drew much criticism from environmentalists and members of the public. Saying "We will mine more, drill more, cut more timber," Watt ramped up clear-cutting on the national forests; significantly expanded coal, oil, and gas leasing; and proposed the sale of 5 percent of the federal public lands.[31]

Reagan's anti-environmental agenda—combined with media coverage of Watt and Gorsuch—precipitated a significant public backlash. The 1980s witnessed a substantial and sustained increase in public support for environmental protection, evidenced by public opinion polls and skyrocketing memberships in environmental organizations. The Wilderness Society,

Audubon Society, and the National Wildlife Federation all reported steep rises in their membership rosters during the 1980s. The Sierra Club's membership grew exponentially in the 1980s; from 1980-1982, the Sierra Club gained 145,000 new members.[32] By 1989, Gallup reported that 76 percent of Americans considered themselves environmentalists.[33] Even more significantly, Gallup reported that from 1985-1990, the percentage of Americans who prioritized protecting the environment over economic growth—a belief central to the new environmental paradigm (NEP)—rose from 61 percent to 71 percent.[34]

The public backlash against the Reagan administration's environmental policies drove home an important lesson for American conservatives: direct attacks on environmental protection would be untenable; they needed a more subtle strategy. They realized they needed to disempower and delegitimize the popular environmental movement. So they began a slow-burning campaign against environmentalism by politicizing and discrediting the movement through stereotypes, slurs, and unfounded accusations about environmentalists. In time, the once popular environmental movement came to be seen as a radical left-wing movement out of touch with ordinary Americans.

From Environmentalists to Tree Huggers

One way of telling the history of the United States is to tell a story about trees. When settlers landed on the eastern seaboard, it is estimated there were about 1 billion acres of forests on the U.S. continent. An adventurous squirrel could have leapt onto a tree in northern Maine and traveled through an unbroken leafy canopy without touching the ground until it reached Texas. As settlers worked their way west, they felled trees, acre by acre, state by state. When settlers reached the Pacific Northwest, they encountered roughly 17 million acres of verdant old-growth fir, hemlock, spruce, and cedars of unimaginable girth and height. When Charles Wilkes, Commander of the U. S. Exploring Expedition of 1838-1842, visited the Pacific Northwest he encountered trees he described as saplings "six feet in diameter and upwards of 200 feet in height. I could not control my astonishment at the size of the trees."[35] An old-growth Douglas Fir can tower almost 200 feet above the forest floor with a trunk almost 13 yards in circumference;[36] the largest known Sitka Spruce has a circumference exceeding 20 yards.[37] By the 1980s, only about three to four million acres of the ancient giants in the Pacific Northwest remained.[38]

Conflict over the fate of the remaining ancient forests of the Pacific
Northwest erupted during the Reagan administration. The administration
launched an accelerated schedule of clear-cutting the old-growth forests.
When you hear the term "clear-cutting," think "mowing." With an assort-
ment of logging machines such as "feller bunchers" and "grapple skidders,"
loggers essentially mowed down the ancient forests. They took everything, leaving behind millions of acres of denuded mountains. Local and
regional environmental groups fought the Reagan administration's aggressive timber harvesting. But in a region dominated by politically powerful
timber interests, environmental activists realized that they needed national attention. One group put a felled old-growth tree on a flatbed truck
and toured the country. Many Americans were horrified to learn that the
Forest Service was auctioning off ancient giants for harvesting; confusing
national forests with national parks, many had no idea the Forest Service
permitted the harvesting of trees on our public lands.

The controversy over the ancient forests in the late 1980s and early
1990s helped broadcast caricatures about environmentalists. What began
as a contested, regional environmental issue grew into a national debate.
President George H. W. Bush accused environmental "extremists" of locking up national resources. Given the national prominence of the issue, caricatures of environmentalists became widely known. The Timber Wars
were fought over preservation of the old-growth cathedral forests. However, in the fight to protect the ancient giants, activists used the most powerful legal tools available: provisions in the National Forest Management
Act and the Endangered Species Act.[39]

Reliance on these potent legal tools had political consequences. Before
long, the public face of the battle became the Northern Spotted Owl, a species dependent upon its old-growth forest habitat. Focusing on the owl enabled opponents of forest protection to trivialize both the issue and the
champions of ancient forest preservation. Instead of focusing on ancient
forests that provide countless ecosystem services to people—the whole
point of the battle over the old-growth forests—attention turned to saving
an owl. Bumper stickers, billboards, t-shirts, and caps sprouted slogans
like "Save a logger, eat an owl!" and "I love Spotted Owls FRIED." Environmentalists were mocked as liberal "tree huggers." Conservative radio
commentator Rush Limbaugh repeatedly told an audience of millions that
environmentalists were "hippie-dip wackos" who cared more about owls
than people.[40]

At the same time, conservatives were able to capitalize on emerging anti-environmental sentiment in the American west as the young environmental movement collided with traditional assumptions about the economic uses of our public lands. Almost half of the American west is federal land. The Forest Service manages more than 166 millions acres in the west; the Bureau of Land Management manages over 245 million acres of western public lands. The Forest Service manages our national forests based on the principle of "multiple use management." According to the Multiple Use Sustained Yield Act, the Forest Service is charged with managing for outdoor recreation, range, timber, watershed, and wildlife and fish purposes.[41] Unfortunately, many of these "multiple uses" are incompatible with each other: it is impossible to protect fish and wildlife and many forms of outdoor recreation while simultaneously managing for timber production in the same geographic area. For example, clear-cutting has compromised miles of prize-winning trout streams and salmon runs in the Pacific Northwest; and most hikers and campers don't want to recreate in or near logged forests. Traditional westerners, accustomed to having access to our public lands for cattle grazing, timber harvesting, and mining, stridently opposed new environmental policies designed to protect ecosystems, species, and potential wilderness areas. Over their objections, there have been reforms to timber harvesting and cattle grazing; however, the law that governs mining on our federal lands has not been revised since 1872.[42]

Resentment over protecting the last remaining old-growth in the Pacific Northwest soon morphed into a quasi-grassroots form of anti-environmentalism known as the Wise Use Movement active in the west. The Wise Use Movement was inspired by the earlier "Sagebrush Rebels" opposed to restrictions on the use of western public lands. While it is true that many timber workers genuinely feared for their job security if environmentalists succeeded in winning the battle to protect the ancient forests, much of the so-called grassroots opposition was in fact manufactured by conservative political entrepreneurs and the timber industry. Ron Arnold, along with Alan Gottlieb (founder of the conservative Center for the Defense of Free Enterprise), launched the Wise Use Movement in 1988. Arnold and Gottlieb bastardized the meaning of the slogan "wise use," coined by the first chief of the Forest Service, Gifford Pinchot. For the Forest Service, "wise use" meant not overusing or underusing natural resources. For the Wise Use Movement, "wise use" meant unrestrained timber harvesting.

In the early days, the Wise Use Movement was essentially an astroturf group—fake grassroots—created by the timber industry opposed to restrictions on logging. The industry gave timber workers paid days off and transportation to go to anti-environmental rallies. The industry played an important role in launching the Wise Use Movement; in time, genuine grassroots anti-environmental sentiment took root. The combination of anti-environmental corporate support, timber workers' economic anxieties, traditional westerners' fierce independence, and caricatures of environmentalists broadcasted by conservative politicians and pundits proved to be a powerful catalyst for anti-environmentalism more broadly. According to Mark Stoddart et al., "the discourse of the Wise Use movement has since been mainstreamed in US political discourse through integration into the Tea Party, as well as through Koch-funded groups like Americans for Prosperity."[43]

Tree Huggers to Woke Radicals

At the same time that environmental tree huggers and "hippie-dip wackos" were advocating for the preservation of our last remaining ancient forests, conservative opponents of environmentalism seized on the fall of the Soviet Union as an opportunity to politicize the environmental movement. Environmentalists became socialist "watermelons," green on the outside, red on the inside. The conservative columnist Charles Krauthammer claimed that "under the banner of socialism . . . social control, once asserted on behalf of the working class, is now asserted on behalf of the spotted owl."[44] Alan Gottlieb, who partnered with Ron Arnold in forming the Wise Use Movement, saw a lucrative fund-raising opportunity. As a highly successful direct mail fundraiser, Gottlieb knew the importance of having an "evil empire" to stoke fear and increase the size of contributors' donations.[45] The end of the Cold War deflated the communist threat; Gottlieb realized the environmental movement could be the new "perfect bogeyman."[46] A related line of attack emerged after the United States participated in the 1992 Earth Summit. Environmental sustainability was equated with one world government under the auspices of the United Nations' Agenda 21.[47] By the end of the decade, environmentalism had been transformed into a left-wing political movement out of touch with ordinary Americans.

In recent years, the environmental movement's attention to environmental and climate justice has offered new opportunities for conservatives to further politicize the movement. After years of regressive climate policy under the first Trump administration, renewed calls for climate justice have dominated the public narrative about climate change and given conservatives another line of attack: the environmental movement became a woke social justice movement.

In 2022, Paul Krugman argued that conservatives have driven a further wedge in public opinion about environmental issues by aligning environmental protection with "woke capitalism," conflating cultural issues of race and identity with environmentalism.[48] The environmental movement's focus on environmental and climate justice paired with public demands were reflected in the Biden administration's climate policy initiatives. The Biden administration integrated climate justice in its overall approach to climate change policy. For example, "Securing Environmental Justice and Spurring Economic Opportunity" is embedded in Executive Order 14008, "Tackling the Climate Crisis at Home and Abroad."[49] Executive Order 14008 launched the Justice40 Initiative, which directs 40 percent of the overall benefits of federal investments in clean energy, affordable housing, and pollution reduction to be directed toward disadvantaged communities.[50] Not surprisingly, a 2023 Pew Research Center poll found a wide partisan gap over the direction of the Biden administration's climate policies. Pew reported that 76 percent of Democrats thought Biden's climate policies were taking the country in the right direction; a mere 8 percent of conservative Republicans and only 27 percent of moderate to liberal Republicans agreed with the direction of Biden's policies.[51]

A Slew of Slurs

According to one observer, "There have been few areas in which right-wing abuse was so fecund as with anti-environmentalism."[52] To this day, one can find a multitude of slurs and caricatures of environmentalists in the media and popular press. Over the years, a stunning list of derogatory labels for environmentalists has emerged. In addition to eco-nuts, eco-freaks, greenies, and eagle freaks, Frederick Buell notes that environmentalists have been called "psychological and social misfits—wooly-headed, sentimentalist, nostalgic purists" as well as "neo-Luddites and neoprimitives."[53] Nadia Bashir et al.'s 2013 study of stereotypes of activists revealed an astonishing list of negative characterizations of environmental activists. Frequently mentioned depictions included hippies, unhygienic, irrational, militant, eccentric, overreactive, self-righteous, hairy, stupid, zealous, nontraditional, and crazy, among others.[54]

Anna Klas et al. found that some members of the public hold positive views about environmental activists, yet at the same time they also have many negative perceptions of them including "aggressive, arrogant, unrealistic, and stubborn" as well as "being egotistical and didactic and overly idealistic."[55] The subterranean influence of the dominant worldview also shapes cultural characterizations of environmentalists. Flowing from the assertion that the ecological crisis is overblown, environmentalists are regularly accused of exaggerating the threats posed by environmental deterioration, and the legitimacy of their views is discounted. For example, in media coverage of his 2020 bestseller *Apocalypse Never: Why Environmental Alarmism Hurts Us All*, Michael Shellenberger apologized for environmental "fearmongering" in his earlier days as an environmental activist. [56]

A 2022 episode of the podcast *Drilled* examines the long-running campaign to caricaturize environmentalists. Host and journalist Amy Westervelt noted, "For decades, the fossil fuel industry has successfully framed environmentalists as silly, elitist, radical and out of touch."[57] Conservative politicians, commentators, and evangelical groups also have contributed many negative characterizations of environmentalists. In 2006, long-serving Republican Representative from Alaska, the late Don Young, described environmentalists as a "self-centered bunch of waffle-stomping, Harvard-graduated intellectual idiots. . . . They are not Americans, never have been Americans, never will be Americans."[58] Over time, Rush Limbaugh added "econazis" and "jihadists" to his lexicon of slurs.[59] In *Apocalypse Never*, Shellenberger, depicted environmentalists as "lost souls" who seek religious transcendence by worshipping "false gods."[60] The Cornwall Alliance, a Christian evangelical group, describes environmentalism as a "native evil." In a series of lectures on DVD and in a book called *Resisting the Green Dragon: Dominion Not Death*,[61] followers are urged to resist the "Green Dragon" of environmentalism.

STALLING THE ENGINE OF SOCIAL CHANGE

Compared to other social movements, environmentalism is unique for its heavy reliance on scientific evidence.[62] Conservatives' attacks on environmentalists have a dual purpose: they hope to discredit environmentalists as a voice of science—shooting the messenger—and they hope to neuter the political effectiveness of the movement. Delegitimizing environmentalism stymies the movement's potential as an engine of political mobilization and social change. Cultural narratives and many media portrayals of environmentalists fall in line with conservatives' campaign of caricatures. As a society, we ignore the important role environmentalists play in society and politics. We do not treat environmentalists as legitimate political advocates, and we rarely equate environmentalists with credible science.

As the most visible face of the environmental movement, the mainstream groups working in Washington are labeled a "special interest." As John Meyer notes, "environmentalists are seen as representing a particular interest—one among many—that any nominally democratic or pluralistic political system should consider when making policy."[63] From the earliest days of the environmental movement, the mainstream's advocacy of nature protection often has been criticized as an elitist argument to protect nature's playground for the affluent.[64] In 1977, William Tucker's article "Environmentalism and the Leisure Class" planted the seeds of the elitist critique of environmentalism.[65] Critics also claim that mainstream groups selectively advocate for policies that enhance their organizational budgets; some mainstream groups have been criticized for accepting corporate donations that influence their organizational agendas.[66]

The realities of organizational survival in the competitive world of non-profit, public interest advocacy are indeed messy. However, like all non-profit organizations, the environmental mainstream must make strategic choices for organizational survival. We also regularly overlook the longevity of the major mainstream groups. The Sierra Club—still active to this day in environmental politics at the federal, state, and often municipal levels—was founded in 1892. As Christopher Bosso emphasizes, "The seeming permanence of organizations that work on behalf of environmental values elicits little comment. . . . It strikes me as odd that so few commentators consider this fact worthy of examination."[67] And perhaps we should be reminded of the mainstream's fundamental purpose and collective organizational mission: environmentalists are working to protect public health and our shared planetary home. Blinded by charges of elitism and organizational self-interest, we overlook the fact that for decades, the mainstream groups have been advocating for the protection of intact, functioning ecosystems, species diversity, and wilderness areas vital to the overall health and stability of the Earth System.

We also ignore the fact that the umbrella of today's environmental movement encompasses the original large mainstream groups, a countless number of smaller, locally based environmental justice groups fighting community environmental health hazards, and young climate activist groups dedicated to climate action and climate justice. The contemporary movement comprises issues ranging from the protection of wilderness and biodiversity to the health hazards of industrialization and even our cookware to justice for displaced fossil fuel workers. The stereotypical white Sierra Club member has been joined by low-income Americans, Blacks, Hispanics, Native Americans, and an energetic and vocal young generation of activists deeply worried about the fate of our planetary home.

In their campaign to undermine the power of the environmental movement, like broadcasting weed seeds in fertile fields, conservatives launched a broadscale attack on environmentalists that continues to this day. Arguably, the campaign to delegitimize environmentalists has been highly effective. Gallup reports that the percentage of Americans who consider themselves environmentalists dropped significantly from 1989 to 2021. In 1989, Gallup found that 76 percent of Americans considered themselves environmentalists. Despite the breadth and diversity of today's environmental movement, by 2021, only 41 percent of Americans considered themselves environmentalists.[68] According to a Pew Research Center poll, even fewer millennials self-identify as environmentalists. In 2014, only 32 percent of millennials identified as environmentalists; many consider environmentalism to have been "corrupted and politicized."[69] Such a precipitous drop in the number of Americans who consider themselves environmentalists is puzzling given recent polling data on global warming. An October 2021 Gallup poll found that 65 percent of Americans worry a great deal or a fair amount about global warming; 43 percent believe that global warming will pose a serious threat in their own lifetime.[70] Researchers with the Yale Program on Climate Change Communication reported a "Dramatic increase in public beliefs and worries about climate change."[71]

One might think that dramatically escalating public worries about climate change would lead Americans to embrace environmentalism. However, a 2023 Pew Research Center survey found that "53% of Americans say they've felt suspicious of the groups and people pushing for action on climate change." Republicans are even more skeptical of climate activists: the same survey found that 78 percent of Republicans are suspicious of climate activists.[72] It is possible that some individuals simply reject the label "environmentalist" while remaining committed to environmental protection; this may be true of millennials, many of whom reject the term as "outdated."[73] Younger generations of climate activists rally around the call for climate justice. It also could be that some people eschew the term "environmentalist" because they do not think their own lifestyle is green enough for them to deserve the term. Klas et al. found that many members of the public believe that to be considered an environmentalist, one must hold biocentric values and engage in a wide range of pro-environmental, individual-level behaviors in their personal lifestyle.

A public narrative focused on personal responsibility has been very dominant in cultural discourse about environmentalism for decades. Years ago, the fossil fuel, plastics, and chemical industries purposefully created a public narrative of personal responsibility to direct attention away from broader systemic changes in manufacturing and energy use.[74] For many, this long-running narrative—distributed via posters, articles, and books such as *Fifty Simple Things You Can Do to Save the Earth*—has created a personal burden as well as a reason to reject personally identifying with the term "environmentalist."[75] The bar appears too high. However, Klas et al. also found that negative perceptions of environmentalists can interfere with people's willingness to adopt environmentally friendly behaviors and lifestyles.[76] Similarly, Bashir et al. report that individuals avoid adopting behaviors advocated by environmental activists because they associate the activists with negative stereotypes, viewing them as "people with whom it would be undesirable to affiliate."[77]

Although complex reasons may drive the declining number of people who self-identify as environmentalists, we cannot ignore the effects of years of political and cultural discourse that denigrates and caricaturizes environmentalists. The sharp decline in the percentage of Americans who consider themselves environmentalists coincides with conservatives' orchestrated campaign against the environmental movement that began in the late 1980s. Pollsters report that a widening partisan gulf between conservatives and liberals particularly around environmental issues and climate change emerged in 1989.[78] Studies by Bashir et al., Klas et al., and others have found that people avoid environmental activists because they link them with negative stereotypes.

The negative characterizations and even outright avoidance of environmentalism have significant social and political consequences. Most environmental scientists, commentators, and environmental activists agree that although individual lifestyle choices are important, broad systemic change is essential for lasting environmental protection. Disturbingly, Klas et al. found that people were more likely to describe environmentalists in negative terms when environmentalists "choose to engage in collective, public sphere behaviors that challenge the status quo and may even destabilize the system."[79] For conservatives cranking out caricatures of environmentalists, their slurs and derogatory labels have achieved the twin goals of preventing more widespread adoption of greener lifestyles among the American public as well as reducing pressure for political change, both of which would contribute to a shift in societal worldviews.

The ongoing effort to politicize environmentalism has consequences. As the public face of the environmental movement, discrediting environmentalists hobbles the movement's effectiveness as an engine of social change. After decades of caricatures and negative labels promulgated by conservatives, it is not surprising that almost half as many Americans call themselves environmentalists compared to thirty years ago. The efficacy of the environmental movement as a driver of social change is severely compromised when many Americans view environmentalists as "people with whom it would be undesirable to affiliate."[80]

To ponder the implications of changing public views of environmentalism, consider a different social movement. In the wake of *Dobbs v. Jackson Women's Health Organization,* what if Americans had denigrated and shied away from the activists who represented the public face of the reproductive choice movement, believing that doctors, public health advocates, and women's groups were "people with whom it would be undesirable to affiliate"?[81] Despite accusations that pro-choice advocates were baby killers and murderers, Americans in red states and blue did not turn away from the reproductive choice movement.

Admittedly, attacks on reproductive choice have very personal consequences. Sadly, the conservative assault on the environmental movement and environmental science also has affected people's lives in very personal ways. Families living in the shadow of polluting industries such as plastics and petroleum refining facilities watch in horror and anguish as family members are stricken with cancers and respiratory and cardiovascular disease. Thousands of American families living in communities regularly shrouded in air pollution watch their asthmatic children struggling to breathe. How many people have lost their homes and even their entire communities in the wake of devastating wildfires and floods turbocharged by global warming? Arguably, conservatives' successful efforts to delegitimize the environmental movement have hampered our societal ability to aggressively protect public health, biodiversity, vital ecosystems, and even the stability of our global climate.

A popular environmental movement does not automatically prevent these environmental, social, and public health tragedies. However, the history of the environmental movement suggests that an energetic, popular social movement makes a difference. Shortly after 20 million people participated in the first Earth Day, Congress passed a record amount of landmark environmental legislation that was signed into law by President Nixon. In 1994—still early in conservatives' attack on environmentalism— Republicans took control of the House of Representatives for the first time in decades. Speaker Newt Gingrich spearheaded the launch of Republican's Contract with America, a set of legislative initiatives to be completed within the first hundred days of the 104th Congress. Sold to the American people as a plan to reduce the federal deficit and regulatory overreach, it also contained a variety of proposals to undermine environmental protection. Mainstream environmental groups such as the NRDC and EDF, working in concert with the media, unearthed anti-environmental proposals buried in the Contract with America. When Americans learned about the anti-environmental reality of the Contract, Republicans faced a public backlash reminiscent of the furor over the Reagan administration's anti-environmental policies. Proposals in the Contract included slashing the EPA budget for environmental enforcement, rolling back regulations governing pesticides in food, and a proposal to weaken the Clean Water Act, among others. The NRDC dubbed the Republicans' plan the "Dirty Water Act" after it was revealed that proposed changes in dozens of programs under the Clean Water Act would have weakened pollution controls.[82] Republicans were not able to enact their agenda of anti-environmental measures.

The Dirty Water Act and Republicans' other anti-environmental measures received much public attention, thanks in large part to the efforts of the mainstream environmental groups in Washington. So Republicans turned to a stealth strategy. Republican legislators attached a precedent-setting number of anti-environmental riders to high stakes spending bills hoping that President Clinton would be forced to sign them into law.[83] Once again, the environmental mainstream and the media came to the rescue, and word of Republicans' stealth strategy got out; The *Washington Post* published an opinion piece called "Riders from Hell."[84] When President Clinton vetoed the spending bill, the era of government shutdowns was born; the federal government shut down for nearly three weeks in 1995-1996.[85] Not surprisingly, when the George W. Bush administration was about to take office, they sought the advice of Republican political strategist Frank Luntz on how Republicans could improve their anti-environmental public image.

Admittedly, much has changed in American politics since the earlier days of the environmental movement. Partisan gerrymandering, voting restrictions, and the Senate filibuster undermine democratic representation and the power of majority public opinion. Today's Supreme Court differs dramatically from the Burger Court sitting during the first Earth Day. In one of the most consequential decisions in decades, in 2024 the Supreme Court dramatically constrained the authority of regulatory agencies like the EPA to regulate in the name of environmental and public health.[86] At the state level, red states have overruled local decisions to improve environmental quality in predominantly blue cities. States like Texas and Idaho have passed state legislation to block cities from adopting climate policies in their city charters or mandating stricter energy efficiency codes.[87] Similarly, conservative legislatures in Arizona, Missouri, Idaho, Wisconsin, Minnesota, Florida, Indiana, and Iowa have enacted their own version of "bag bans," blocking cities from passing bans on single-use plastic bags.[88]

Political restructuring and minority decisions that thwart the will of the majority are compounded by extreme polarization among the American public. Yet Americans always have had legitimate political differences over a host of issues including gun rights, reproductive choice, immigration, taxation, the role and size of government, and many others. This is the norm in democratic politics. However, the partisan divide over environmental protection was artificially created by conservatives to thwart the power of the environmental movement. Nearly 80 percent of Republicans self-identified as environmentalists shortly after the Reagan administration tried to eviscerate environmental protection. Republicans' identification with the environmental movement began a downward spiral in 1989 coincidental with conservatives' war on the environmental movement. By 2021, only 25 percent of Republicans self-identified as environmentalists.[89] Bipartisan political advocacy of stronger environmental legislation ended with the George H. W. Bush administration just as caricatures and stereotypes of environmentalists entered the cultural bloodstream.

Resurrecting a bipartisan environmental movement—challenging the cultural narrative about environmental extremists and wackos—is essential to the health of our planet and our democracy. Although Americans of different political stripes often have profoundly conflicting views on what social, cultural, and political life in America should look like, we are united by our shared residency on the North American continent, on this single blue ball in space. The constant chyrons of partisan messaging obscure the enduring reality of Americans' love of our natural heritage. Our love of the land—for "amber waves of grain to purple mountain majesties"—could reunite Americans all across the country. In 2012, a bipartisan team of pollsters reported that "conserving the country's natural resources—land, air and water—is patriotic."[90] We live in a dangerously warming world of rampant ecological degradation awash in plastics and toxic forever chemicals, a world of wounds oblivious to political affiliation. Attacking environmentalists and weakening the environmental movement has personal and political consequences for us all.

FOUR

THE SIEGE ON

ENVIRONMENTAL SCIENCE

For decades, conservative elites have been engaged in a two-pronged assault on the environmental movement. They have caricaturized and stereotyped environmentalists with derogatory labels and unfounded accusations to discredit the movement in the eyes of the public. They have politicized the environmental movement by characterizing it as a radical, left-wing political movement. In a parallel attack, conservatives have denied, distorted, and fabricated environmental science to further undercut the science-based environmental movement and to prevent a policy response to environmental decline and climate change. Both legislative and regulatory deliberations depend heavily on scientific assessments of the threats to human welfare and environmental well-being. Manufacturing doubt about the accuracy and authority of environmental science undermines public demands for environmental protection and further weakens the environmental movement. Conservatives also have politicized environmental science by equating science with policy, thus discrediting the authority of scientific opinion. Politicizing the science reinforces the notion that the environmental movement is a left-wing political movement using fake science to promote its own vested interests.

Attacking environmental science also is an important strategy to counter the momentum of the environmental movement and block a po-

78

tential shift in societal worldviews. The political force of the environmental movement is driven by both science and public support. A vast body of environmental science tells us that the dominant worldview offers a mythological view of humans and nature totally at odds with ecological and planetary realities. A scientific paradigm shift can occur when "a significant body of knowledge or information accumulates that is contradictory to, or unexplained by, the accepted paradigm."[1] The body of scientific evidence about planetary decline is ramming the fortress walls of the dominant paradigm. However, the weight of scientific evidence alone isn't enough to drive a social paradigm shift. The scientific argument must be carried forward by citizens—and voters—working together to force social, political, and environmental change. The power of a bipartisan popular environmental movement paired with the truly staggering amount of scientific evidence of planetary decline could drive a shift in societal worldviews.

The previous chapter discussed conservatives' war on the environmental movement. This chapter examines the parallel strategy in conservatives' campaign to disempower the movement and obstruct policy initiatives: a decades-long campaign of scientific distortion and deceit paired with attacks on environmental scientists. The chapter concludes with a look at extinction denial, the next chapter of environmental science denial.

THE EARLY YEARS

Early in his administration, President Reagan paved the way for the full-fledged campaign against environmental science. Soon after taking office, Reagan was faced with the prospect of an international treaty with Canada initiated during the Carter administration to address acid rain. Reagan challenged the science, even though the scientific consensus on the causes and effects of acid rain were well established. Arguing that the science was not settled, Reagan commissioned a decade-long study of acid rain to delay action. Soon the public backlash against the Reagan administration's environmental policies drove home an important lesson. Conservatives could not overtly challenge environmental protection. They needed a more devious strategy. Just as conservatives set their sights on undermining the environmental movement, in a parallel vein they realized that undermining the science could help them achieve their political and policy goals without appearing to be anti-environmental. For a public audience, they could emphasize the need for sound science and careful scientific reviews while making hollow statements about the importance of environmental protection. By attacking the science, and the "voices" of science, conservatives could contest the severity of environmental problems, undercut the case for environmental regulation, and more fundamentally, protect the DSP worldview. As Aaron McCright and Riley Dunlap observed, "Conservatives go to great lengths to mask their efforts to weaken environmental protection by attacking the scientific evidence concerning environmental problems."[2]

Enablers and Early Tactics

Chief among conservatives' early enablers were a handful of influential scientists dubbed the "merchants of doubt."[3] Prominent physicists such as Frederick Seitz and Fred Singer had much to offer conservatives in their campaign against environmental science. Seitz and Singer had cut their teeth on the art of casting doubt on credible science by working with the tobacco industry; they helped the industry challenge the scientific evidence on the health effects of smoking. Familiar names in Washington circles, and experts in manufacturing doubt and challenging the integrity of credible science and scientists, Seitz and Singer (and others) were ideal partners for conservatives seeking to undermine the still young environmental movement by attacking the science. Singer became a favorite among conservative think tanks for his vigorous attacks on climate change science.

The "merchants of doubt" showed conservatives the power of language. Manipulating language became an important weapon in conservatives' campaign against environmental science. Throughout the 1990s and continuing into the George W. Bush administration, conservatives frequently disparaged "junk science." Rather than directly challenging the science of specific environmental issues, they regularly proclaimed that environmental policies needed to be based on "sound science." The focus on "junk science" and "sound science" can be traced to the work of The Advancement of Sound Science Coalition (TASSC), founded in 1993 to advocate for "sound science" in policymaking. Although TASSC was created as a front group for the tobacco industry to challenge the EPA's focus on secondhand smoke, like the original "merchants of doubt," soon TASSC turned their attention to environmental science. "Junk science" was described as "bad science used by environmental Chicken Littles to fuel wacky social agendas."[4] What exactly is junk science? The work of environmental scientists working in academia and prestigious scientific institutes became junk science. More importantly, where was sound science to be found? Scientists from industry and conservative think tanks were deemed trustworthy sources of sound science.

"Sound science" was bandied about frequently during the George W. Bush administration. In his advice to the Bush administration, Republican political consultant Frank Luntz advised the administration to focus on sound science, emphasizing "Sound science must be our guide in choosing which problems to tackle and how to approach them."[5] References to sound science played a starring role in conservatives' ongoing efforts to weaken the Endangered Species Act, arguably one of our most powerful environmental laws designed to prevent the irreversible loss of species and their genetic inheritance. For example, the 107th Congress released H.R. 4840, the Sound Science for Endangered Species Act Planning Act of 2002. The language of the act hints at the meaning of sound science and the proponents' intentions:

> The Sound Science for Endangered Species Act Planning Act of 2002 amends the Endangered Species Act of 1973 to require the use of the best scientific and commercial data available as a basis of determinations on a petition to add or remove a species from the endangered species list. Directs the Secretary of the Interior to give greater weight to any scientific or commercial study or other information that is empirical or has been field-tested or peer-reviewed.[6]

The term "sound science" had obvious appeal for members of the public. On the surface, one might think the Sound Science for Endangered Species Act would have strengthened the scientific basis for species' protection. However, members of the public never knew what "sound science" actually meant. For those knowledgeable about endangered species policy, references to "commercial data" and "commercial study" made the goals of the Act clear: decisions based on "sound science" relied on data provided by commercial interests rather than "junk science" from biologists working in academia and government. Conservatives also manipulated the meaning of peer review to stymie agency decisions unpopular with industry by creating cumbersome, time-consuming processes. In other words, peer review meant that federal agencies needed to utilize industry-friendly plans that required such exhaustive analyses that agencies were prevented from taking prompt action to protect public health and the environment.

Ask the Economists and Political Scientists

As the conservative campaign to delegitimize environmental science evolved, challenges to the science of specific environmental issues became a central tactic. Conservative legal scholars, economists, and policy analysts with no training in the environmental sciences—often working with conservative think tanks such as the Competitive Enterprise Institute, the Cato Institute, the Heartland Institute, and others—have produced a steady stream of pseudo-scientific publications including books, editorials aimed at the public, and policy briefs for journalists and policymakers.[7] These specious reports are based on cherry-picked statistical data to challenge the scientific evidence on multiple issues including acid precipitation, the ozone hole, biodiversity loss, and global warming. These scientifically meaningless exercises in selective number crunching—presented as legitimate alternative perspectives on environmental problems—gain traction in policy circles and the media, distort scientific reality, cast doubt on the work of credible environmental scientists, and confuse the public. To make matters worse, many of these reports are printed in formats that mimic genuine peer-reviewed studies.[8] Unfortunately, representatives from these think tanks are treated as credible experts by the media and often sought out as sources of expert opinion.[9]

Bjørn Lomborg's 2001 book *The Skeptical Environmentalist: Measuring the Real State of the World* offers a classic early example of the distortion of environmental science.[10] In this hefty volume, Lomborg—a Danish political scientist with expertise in statistics and game theory—set out to re-interpret the scientific data across a broad swath of environmental disciplines and topics. In 540 pages of cherry-picked statistics, Lomborg challenged the established science on forests, energy, water resources, air and water pollution, toxic pollution, acid rain, biodiversity, and global warming. Lomborg has no training in the biophysical sciences; not surprisingly, *The Skeptical Environmentalist* was roundly criticized by environmental scientists. Nevertheless, his book was greeted with enthusiasm by a wide variety of conservative commentators. A review of the book by the *Economist* stated, "This is one of the most valuable books on public policy—not merely on environmental policy—to have been written for the intelligent general reader in the past ten years. *The Skeptical Environmentalist* is a triumph."[11]

Similarly, the *Wall Street Journal* described Lomborg's book as "a superbly documented and readable book."[12] Many members of the public probably were relieved by Lomborg's misleading claims about the severity of our environment problems. The average citizen does not understand that statistical dexterity is insufficient to analyze complex environmental data. More insidiously, competing so-called scientific assessments of environmental problems written by conservative non-scientists convey the notion that scientific knowledge is contested and uncertain, further confusing the public.

AN ECOSYSTEM OF CLIMATE SCIENCE DENIAL

The conservative campaign to rebut and politicize environmental science went into high gear with the advent of global warming as a public issue in the late 1980s. The magnitude of the environmental, political, and economic threats and stakes associated with global warming has led to the creation of an entire conservative ecosystem of global warming denial designed to protect political and economic fortunes grounded in our dominant societal worldview. Moving away from fossil fuels represents an obvious threat to those whose political and economic fortunes are tied to coal, oil, and gas. On a deeper level, ending the fossil fuel era represents a major threat to defenders of the dominant worldview, a worldview intimately tied to a free-market economy powered by cheap fossil fuels and unrestrained resource extraction. Over the years, the conservative campaign to politicize and manipulate climate change science has evolved from more blatant science denial and distortion into a dangerous hydra of devious, nuanced tactics. Beginning with the early activities of the fossil fuel industry, this section tells a chronological story of conservatives' efforts to refute the reality of global warming by challenging and undermining climate change science, framing climate change as a liberal cause, and discrediting scientists.

Fossil Fuel Industry Fabrications

The fossil fuel industry orchestrated a carefully crafted media strategy to distort the realities of climate science and plant seeds of doubt in the minds of average Americans that has spanned decades. The industry created a playbook for conservatives to replicate in their broader campaign against environmental science. Learning from the original "merchants of doubt," fossil fuel companies strategically manufactured scientific uncertainty to cast doubt on the credibility of global warming science.

Fossil fuel industries had an early interest in global warming. They realized decades ago that burning fossil fuels would produce greenhouse gases that could warm the global climate. In 2015, *Inside Climate News* revealed a powerful story based on ExxonMobil's internal documents and records.[13] *Inside Climate News* found that in 1979, Steve Knisely, an intern at Exxon Research and Engineering, was asked to analyze how global warming might affect use of fossil fuels. Knisely projected that there would be "noticeable temperature changes" and 400 parts per million (ppm) of carbon dioxide (CO_2) in the air by 2010 if fossil fuel use was not constrained.[14] Atmospheric CO_2 hit 400 ppm in May 2013, three years after the Exxon intern's prediction.

Throughout the 1980s, Exxon researchers worked with university and government climate scientists to develop climate models. Company records showed that Exxon had confirmed the scientific consensus on anthropogenic global warming with their own in-house climate models by 1982. Exxon was not the only oil giant aware of the realities of global warming in the 1980s. Shell Oil published an internal report in 1988 called "The Greenhouse Effect" documenting the impacts of global warming on their industry and the world. Shell's confidential report warned of significant changes in sea levels, ocean currents, and precipitation patterns that could affect living standards and food supplies. They wrote, "by the time the global warming becomes detectable it could be too late to take effective countermeasures to reduce or even to stabilize the situation."[15]

Yet, beginning in 1989, Exxon (which became ExxonMobil in 1999) systematically set out to manufacture scientific uncertainty to confuse and manipulate the public.[16] ExxonMobil leaders publicly disparaged their own scientists' work. Although the climate models were becoming increasingly more powerful and reliable, ExxonMobil argued that "uncertainty inherent in computer models makes them useless for important policy decisions."[17] Why the change in company policy in1989? In June 1988, James Hansen, a NASA climate researcher, testified before Congress, affirming, "Global warming is real and happening now."[18] Hansen's testimony could have triggered a policy response. By fueling the notion that the science on global warming was not settled, ExxonMobil and political conservatives hoped to prevent or delay legislative and regulatory measures to combat global warming.

Responding to the 2015 *Inside Climate News* story, officials from Exxon-Mobil rejected the allegations that they had deliberately misled the public and laid down the gauntlet: "Go ahead, you really should. Read the documents *Inside Climate News* cites that purportedly prove some conspiracy on ExxonMobil's part to hide our climate science findings."[19] Geoffrey Supran and Naomi Oreskes took up the public challenge and found even more evidence of ExxonMobil's public deception.[20] They found that early on, ExxonMobil had adopted a three-pronged communications strategy. Their peer-reviewed research confirmed that anthropogenic global warming (AGW) was "real, human-caused, serious, and solvable, while recognizing uncertainties"; their internal business documents acknowledged "the business threat and uncertainties" of global warming.

In sharp contrast, ExxonMobil's paid "advertorials" (also called op-ads) "overwhelmingly expressed doubt that AGW is real, human-caused, serious, or solvable." For fifteen years (1985-2000), ExxonMobil published advertorials every Thursday in the *New York Times*.[21] The advertorials had titles such as "Lies They Tell Our Children," "Apocalypse No," and "Who Told You the Earth Was Warming . . . Chicken Little?"[22] Mobil also published their op-ads in other major newspapers, blanketing the country with lies and disinformation about climate change.

Public utilities joined the major oil companies in a chorus of global warming denial and manufactured uncertainty. A 2018 report released by the Energy and Policy Institute documents public utilities' complicity in hiding the scientific consensus and purposefully sowing seeds of doubt, despite being warned by scientists about global warming as early as 1968. By 1988, the utilities' own research and development organization had acknowledged the scientific consensus on human-caused global warming. However, utilities joined the broader efforts by the major oil companies to "sow doubt about climate science and block legal limits on carbon dioxide emissions from power plants."[23] The lack of infrastructure to support expanded reliance on renewable fuels is frequently cited as a major obstacle to phasing out fossil fuels in our power supply. The utilities chose complicity in a campaign of denial that has hampered our nation's ability to make the necessary energy transition.

Despite efforts by the fossil fuel industries and their conservative allies to ward off a domestic policy response to global warming, international initiatives to protect the global climate proceeded unimpeded. The first international treaty to address climate change, the Kyoto Protocol, was adopted in 1997.[24] The Kyoto Protocol committed industrialized nations to limit and reduce greenhouse gases. Although the United States refused to sign the treaty, the Kyoto Protocol generated political momentum for addressing global warming. The threat of a domestic policy response in lieu of a formal treaty like the Kyoto Protocol focused denialists' strategic response.

The fossil fuel industry's media campaigns were aided by the creation of the Information Council on the Environment, created in 1991 by a coalition of U.S. coal companies and the Edison Electric Institute. Like other astroturf groups at the time, the Information Council also was known as Informed Citizens for the Environment. It was created to "reposition global warming as theory (not fact)."[25] The American Petroleum Institute's Global Climate Science Communications Team also proved to be a strategic partner. The Communications Team initially had used economic arguments to oppose climate change regulations, a staple of conservative politics. After the adoption of the Kyoto Protocol, the Communications Team realized it had been a strategic mistake to argue the economics of regulation rather than challenging the science, writing "industry and its partners ceded the science and fought on the economic issues."[26] Like their conservative political allies, the Communications Team realized that the most effective opposition to regulation lay with challenging climate science.

In 1998, the Communications Team drafted a report outlining a plan to systematically confuse the American public by sowing doubt about climate science.[27] Their Action Plan declared, "Victory will be achieved when . . . average citizens 'understand' (recognize) uncertainties in climate science; recognition of uncertainties becomes part of the 'conventional wisdom.'" The Action Plan enlisted the media in this campaign of manipulation also, saying, "Victory will be achieved when . . . media 'understands' (recognizes) uncertainties in climate science [and when] media coverage reflects balance on climate science and recognition of the validity of viewpoints that challenge the current 'conventional wisdom.'"[28] The American media would go on to become a major purveyor of climate change denial and politicization of climate science. In the name of journalistic balance, the media repeatedly has reinforced the false notion that there are two sides in the climate issue. With the help of the media, climate change became a debate.

Partners in Denial

The fossil fuel industry created a blueprint for climate change denial that was embraced by political conservatives in and outside of government. Over the years, a variety of conservative politicians have issued a steady stream of climate change denial statements. One of the most famous denialists, the late Senator James Inhofe from the oil-rich state of Oklahoma, repeatedly challenged the reality of global warming. Inhofe's oft-repeated declaration that "climate change is a hoax" and his infamous appearance on the Senate floor with a snowball in hand to prove global warming is a hoax have received widespread media attention. Donald Trump proved to be—and continues to be—a powerful bullhorn of climate change denial. Beyond tweeting that climate change is a "hoax invented by the Chinese," both Trump administrations deleted public climate change information and scientific documents about climate change from numerous federal government websites and distorted climate science in various agency reports. Trump remains an influential megaphone of climate change denialism. A 2024 study found that Trump's tweets were the most influential among the universe of tweeting denialists.[29]

Conservative politicians have been aided by conservative political consultants, public relations firms, think tanks, economists, talk show hosts, and contrarian scientists. For years, conservative economists have offered baseless arguments to undermine the case for global warming policy. Economists funded by the fossil fuel industry consistently have grossly overestimated the costs of addressing global warming while enthusiastically underestimating the costs of inaction. Given the primacy of cost/benefit analyses in environmental policymaking, conservative economists have played a key role in undermining the case for regulating carbon in countless climate policy initiatives. According to Benjamin Franta, "The economists used models that inflated predicted costs while ignoring policy benefits, and their results were often portrayed to the public as independent rather than industry-sponsored."[30] Public relations firms working for oil and gas companies and utilities also helped shape the public narrative about climate change by attacking environmentalists, promoting the benefits of the continued use of fossil fuels, while also adding more voices of scientific misinformation.[31]

Political consultants have played a key role in the campaign of denial. One prominent example involves the Republican consultant, Frank Luntz.

In 2001, Luntz wrote a report advising the Bush administration on how to handle its "environmental problem."[32] Luntz offered strategic advice on the global warming issue. The infamous Luntz memo states,

> The scientific debate is closing [against us] but not yet closed. There is still a window of opportunity to challenge the science.... Voters believe that there is no consensus about global warming within the scientific community. Should the public come to believe that the scientific issues are settled, their views about global warming will change accordingly. Therefore, you need to continue to make the lack of scientific certainty a primary issue in the debate.[33]

Luntz's message was clear: we cannot let voters understand the scientific consensus on climate change. Luntz's advice for the Bush administration—in concert with the fossil fuel industries' long-running campaign of distortion—helped solidify the claim that global warming science was not a settled issue within the scientific community, thus adding more uncertainty to the public narrative about climate change.

Conservative think tanks have played—and continue to play—a starring role in the denial and distortion of climate change science. With funding from the fossil fuel industry, the Koch brothers, and dark money, conservative think tanks have produced a steady stream of pseudoscientific papers often written by contrarian scientists like Fred Singer, one of the original "merchants of doubt," aimed at policymakers and the public that challenge the legitimacy of climate science, climate scientists, and the reality of the scientific consensus. For example, the Heartland Institute disputed the scientific consensus for years. The American Enterprise Institute's "A Citizen's Guide to Climate Change" published in 2019 disputes the well-documented acceleration of global warming and claims that rising CO_2 levels have ecological and human health benefits.[34] The Cato Institute's 2016 book *Lukewarming: The New Climate Science That Changes Everything* explains the "real science" of global warming, arguing that "Global warming is not hot—it's lukewarm.... Climate change is real, it is partially man-made, but it is clearer than ever that its impact has been exaggerated."[35]

Conservative think tanks also have aimed their sights on climate change education. Although most think tank publications are directed at political actors, media sources, and the general public, in 2015 the Heartland Institute distributed unsolicited copies of *Why Scientists Disagree About Global*

Warming[36]—written by prominent climate change deniers—to thousands of K-12 teachers and many college instructors. The book has been totally debunked by climate scientists and rebuked by many teachers. Undaunted, the Heartland Institute issued a second edition. More recently, conservatives have supported PragerU with a plethora of climate change denial talking points. PragerU has produced a menu of YouTube videos for schoolchildren with titles such as "Why You Should Love Fossil Fuels," "Can We Really Rely on Wind and Solar Energy," and "Do 97% of Climate Scientists Really Agree?"[37] The videos feature bachelor of arts philosophy major and self-proclaimed "energy expert" Alex Epstein, author of *The Moral Case for Fossil Fuels* and *Fossil Future: Why Global Human Flourishing Requires More Oil, Coal, and Natural Gas—Not Less.*[38]

Perhaps the most powerful weapon in conservatives' arsenal of tactics involves their framing of the public narrative about climate change. Over time, relatively straightforward challenges to the reality of global warming and climate science evolved into the overt politicization of climate change. Conservatives masterfully framed climate change as a liberal political position. By equating climate change science with climate change policy, and emphasizing liberals' acceptance of climate change science, they politicized the issue and created a partisan divide over environmental issues. Liberals' acknowledgment of the scientific consensus on climate change and support for climate change policy became "evidence" of the partisan nature of climate change. A study by Riley Dunlap et al. showed that Republican voters' rejection of climate science is shaped by cues from Democratic elites who support the reality of global warming. Dunlap et al. argue that "Political analysts attribute growing partisan polarization among both political elites and the general public . . . with Republicans becoming increasingly conservative and Democrats increasingly liberal."[39] What is the "increasingly liberal" position? The acceptance of physical reality and the scientific consensus on global warming.

Conservative politicians, political pundits, and talk show hosts helped to characterize climate science as a liberal political agenda. In one broadcast, Rush Limbaugh referred to the polar vortex as "an invention of the liberal Left to further promote the global warming agenda."[40] Conservatives made hay with former Vice President Al Gore's longtime concerns about global warming. In 1992, Gore published a book about global warming called *Earth in the Balance: Ecology and the Human Spirit;*[41] in 2006 he produced the film *An Inconvenient Truth*. Gore's focus on global warming became definitive proof that global warming was a liberal cause. Global warming became "Al Gore's Global Warming." The conservative *National Review* added their own ammunition via their *Planet Gore* blog.[42] Former talk show host Glenn Beck promoted his own book, *An Inconvenient Book: Real Solutions to the World's Biggest Problems,* by telling readers to use his book as "kryptonite against your Gore-worshiping psycho friends."[43] Conservative think tanks also have played a role in politicizing climate science. In 2019, the American Enterprise Institute published a blog post that outlined a blueprint for arguing that the scientific consensus on anthropogenic global warming is merely "politics disguised as science."[44] In other words, the well-documented scientific consensus on human-caused global warming became merely a political agreement.

Conservatives' reactions to the release of the congressional resolution known as the Green New Deal in 2019 further bolstered the public narrative of climate change as a liberal social cause. The Green New Deal called for steep reductions in greenhouse gas emissions, assistance for displaced fossil fuel workers, job opportunities in clean energy industries, and other measures to reduce income inequality.[45] Initially, large numbers of Americans across the political spectrum supported the goals of the Green New Deal. Once conservatives began attacking the Green New Deal, claiming it was a left-wing plot to ban hamburgers, airplane travel, and limit gasoline-powered car ownership—Fox News ran more stories about it than CNN and MSNBC combined—the partisan gap over support for the plan widened considerably, fortifying the notion of climate change as a radical leftist agenda.[46]

Evolving Tactics

As the evidence of climate change becomes more vivid, and the lived experience of average citizens coping with climate change impacts becomes more challenging and disruptive—think massive wildfires and unbreathable air, widespread flooding, droughts, and heat waves—it is becoming untenable, even absurd, to claim that climate change is not real. Paired with the vivid impacts of climate change, the compelling reality of the scientific consensus also has become more difficult to refute. Thus, denialists had to adapt their tactics. A recent form of denial comes from "lukewarmers" who pervert the public conversation about global warming by accepting the reality of global warming but denying the severity of climate change. Björn Lomborg, the "Skeptical Environmentalist," has written two books promoting lukewarmism. He followed his 2007 book, *Cool It: The Skeptical Environmentalist's Guide to Global Warming*, with a 2020 book, *False Alarm: How Climate Change Panic Costs Us Trillions, Hurts the Poor, and Fails to Fix the Planet*.[47] Lomborg's books and others like it join the growing ranks of pseudo-science publications by lukewarmers who acknowledge that global warming is real and mostly human-caused but argue that the threat of climate change has been exaggerated. Conservative columnists also have contributed to the lukewarmers' arguments. George Will wrote an opinion piece published in the *Washington Post* shortly after the release of the IPCC's Sixth Assessment Report in August 2021 refuting the gravity of global warming. He claimed, "With a closer look, certainty about the 'existential' climate threat melts away."[48]

Climate scientist Michael Mann describes the new breed of climate change deniers as "deceivers and dissemblers, namely downplayers, deflectors, dividers, delayers, and doomers."[49] One of the most insidious forms of deflection is the fossil fuel industry's longstanding tactic of shifting blame to individual consumers. Supran and Oreskes analyzed language used by ExxonMobil to shape public discourse around global warming. Instead of taking responsibility for their contributions to dangerous climate change, ExxonMobil's climate change communications have systematically shifted the blame to consumers.[50] Consumers are encouraged to Save the Earth! Working with a public relations firm, British Petroleum created the first carbon footprint calculator, urging consumers to "Reduce your carbon footprint!"[51] However, even the most carbon-neutral lifestyles—complete with state-of-the-art insulation, multi-pane windows, solar panels, electric cars, a vegetarian diet, and the most energy-efficient appliances, affordable only for the affluent—barely make a dent in total carbon emissions. Focusing on individuals takes the spotlight off the real offenders. According to a 2020 report, a hundred fossil fuel companies including ExxonMobil and Shell are responsible for 71 percent of all greenhouse gas emissions.[52] Doomers cap off the manufactured debate by stating that it is already too late to avert global warming; thus changes to the American energy system are too late and unnecessary.

Ever ingenious, climate deniers also have capitalized on the COVID-19 pandemic to spread conspiracy theories about mandatory climate lockdowns. Well-known denialist Marc Morano appeared on Fox News and proclaimed, "We Will Go from COVID Lockdowns to 'Climate Lockdowns' Under Biden."[53] On a totally different front, the plummeting price of solar and wind energy has made renewables affordable for many more people and communities. In response, activists supported by the fossil fuel industry have shown up in midwestern counties to shout down renewable energy supporters at public meetings convened to discuss proposed solar and wind power installations.[54]

The Social Media Megaphone

Members of the public explore climate change amidst the insistent clamor of climate change denial, doubt, and deflection that permeates traditional media, our politics, online search engines, and, increasingly, social media. Conservatives have found popular online platforms to be hospitable homes for climate change denial. The rapid-fire world of social media provides a powerful conduit for amplifying climate change denial and distortion. Emma Bloomfield and Denise Tillery report that climate change denial information "zooms" through online spaces they dubbed the "climate denial blogosphere"; these distortions of actual science are snatched up and "recirculated by conservative media."[55]

The *New York Times* reports "Google's search page has become an especially contentious battleground between those who seek to educate the public on the established climate science and those who reject it."[56] YouTube has been characterized as an "accomplice in efforts by denialists to cast doubt on science."[57] An exploratory study of YouTube videos found that the majority of the videos in the sample opposed the scientific consensus on global warming. The study revealed that "16 videos deny anthropogenic climate change and 91 videos in the sample propagate straightforward conspiracy theories about climate engineering and climate change."[58] The videos in the sample that opposed the scientific consensus on climate change received 16,939,655 views.

Facebook, Twitter, and now X have come under fire for their role in magnifying climate change denial on their platforms. A 2021 report published by the Center for Countering Digital Hate found that 69 percent of climate change denial is found on ten Facebook pages, dubbed the Toxic Ten, maintained by far-right political actors. According to the report, the Toxic Ten are "part of an efficient climate denial disinformation industry, reaching 186 million followers on mainstream social media platforms."[59] The Toxic Ten generated more than $5.3 million in Google ad revenues in just six months. In testimony submitted to the Securities and Exchange Commission in 2022, whistleblower Francis Haugen alleged that Facebook misled investors about their efforts to combat climate change misinformation.[60] In 2023, a report published by the Climate Action Against Disinformation project listed Twitter as the worst purveyor of climate misinformation among the major tech platforms.[61]

Facebook also has helped politicize global warming by classifying "clean energy" and "climate change" as political rather than scientific topics. This designation has affected the posting of educational videos designed to refute false claims about global warming. One of the nation's most prominent climate change scientists and highly acclaimed climate change communicators, Katherine Hayhoe, was discouraged by Facebook's political categorization of climate change. Facebook's political designation would have forced Hayhoe to register her educational videos along with personal information that could have opened her up to personal attacks, a standard tactic of the denialists.[62]

DISCREDITING SCIENTISTS

Inspired by the original "merchants of doubt," conservative politicians, public commentators, and think tanks also have targeted scientists, important "voices" of science. Undermining the professional reputation of prominent climate scientists helps to delegitimize the science and feeds a public narrative of scientific fraud. Conservatives have used a variety of strategies including ad hominem attacks, harassing lawsuits, and death threats to discredit scientists. Although most efforts to discredit environmental scientists are focused on climate scientists, researchers in other fields are not immune. When UC Berkeley biologist Tyrone Hayes's research on the widely used pesticide Atrazine revealed that it was an artificial estrogen causing endocrine disruption in amphibians—male frogs exposed to Atrazine were literally turning into females—Syngenta, the maker of Atrazine, developed a plan to discredit Hayes as a scientist; intimidation came soon after.[63] Hayes received threats of lynching prior to public talks about his research. Even worse, his wife and daughter were threatened with sexual violence.[64]

One of our most prominent climate scientists, Michael Mann, became a target after his research revealed the anthropogenic acceleration of global warming, now known as the famous "hockey stick" graph.[65] Mann has received countless death threats, once opening a letter sent to his office that contained white powder (it turned out to be cornstarch). Speaking to a reporter from the *Washington Post*, Mann said, "I've faced hostile investigations by politicians, demands for me to be fired from my job, threats against my life and even threats against my family."[66] In 2010, the attorney general of Virginia and prominent conservative Ken Cuccinelli claimed Mann's landmark research on the human-caused acceleration of global warming at the University of Virginia represented fraudulent use of taxpayer dollars.[67] Spurred on by Cuccinelli's lawsuit, conservative Representative Joe Barton (R-TX) demanded that Mann and his colleagues turn over fifteen years' worth of research materials and data. Barton then orchestrated a congressional hearing intended to intimidate Mann and other scientists by inviting members of the conservative Competitive Enterprise Institute and the Marshall Institute to lob unfounded accusations at the scientists. Even worse, in 2013, an adjunct scholar affiliated with the Competitive Enterprise Institute published a blog accusing Mann of manipulating and "molesting" climate change data, comparing him to convicted sex abuser and retired Penn State football coach Jerry Sandusky.[68] (Mann filed a defamation lawsuit; in 2024, a jury awarded him $1 million in damages.[69])

Inspired by the infamous Climategate of 2009, demanding years' worth of climate scientists' emails has become a standard weapon in the conservative arsenal for harassing and discrediting scientists.[70] Climategate was a media scandal in which a hacker released thousands of emails from the University of East Anglia's Climate Research Unit—spanning thirteen years of correspondence—to discredit the scientists' work and as fodder for the conservative denial machine. Today, conservative nonprofits use freedom of information laws to force climate scientists to turn over thousands of emails in hopes of exposing them to public ridicule and censure. Demanding thousands of emails also is intended to take valuable time away from scientists' research.

The Energy & Environment Legal Institute has a track record of exploiting open records laws to seek publicly funded scientists' email correspondence. The group's work has been characterized as filing nuisance suits to hamper and disrupt legitimate academic research as part of its broader effort to persuade the public that anthropogenic global warming is a scientific fraud.[71] Harassment of climate scientists is now so widespread that many climate scientists need legal assistance; the Climate Science Legal Defense Fund was created to assist climate scientists who are targeted. The Sabin Center for Climate Change Law at Columbia University and the Climate Science Legal Defense Fund created the "Silencing Scientists Tracker," designed to document reports of efforts to "silence science" by federal, state, and local governments after the 2016 presidential election.[72] Silencing the voices of science is yet another tactic to check the power of environmental science.

THE NEXT CHAPTER: EXTINCTION DENIAL

The focus on global warming in politics and media has overshadowed the deeper, long-running story of conservatives' campaign to delegitimize all environmental science to block social and political change. Efforts to discredit environmental science continue unabated. Although biodiversity typically receives less media coverage than climate change, with increasing coverage of the growing threat of biodiversity loss, we can expect to see the extinction denial movement as the next chapter in this saga.

The unprecedented scale of human-caused biodiversity loss is becoming a planetary emergency. In 2019, the UN's Intergovernmental Science-Policy Platform on Biodiversity and Ecosystem Services (IPBES) released a report emphasizing the unprecedented scope and rate of species loss, stating: "The biosphere, upon which humanity as a whole depends, is being altered to an unparalleled degree across all spatial scales. Biodiversity ... is declining faster than at any time in human history."[73] Charismatic mammals such as tigers and pandas often are the face of the extinction crisis. However, biodiversity loss also includes countless other organisms including amphibians, mollusks, and insects. Warning of an "insect apocalypse," biologists describe insects as the "fabric tethering together every freshwater and terrestrial ecosystem across the planet."[74] Plummeting insect populations underscore the fact that biodiversity loss compromises ecosystem services vital to humans as well as the functioning of ecosystems essential to the stability of larger planetary systems. Unchecked biodiversity loss represents a serious threat to the stability of the overall Earth System; biodiversity loss and climate change are intimately interconnected. Healthy, diverse ecosystems on land and in water help to buffer global warming by absorbing and sequestering carbon. We are already far beyond the zone of safety for both biodiversity loss and climate change, moving perilously close to dangerous tipping points in the Earth System.

According to some analysts, "Biodiversity loss has finally got political."[75] The emergence of the Extinction Rebellion in 2018 with its reliance on controversial protest tactics has raised the profile of the extinction crisis. In 2019, *Scientific American* published an article called "Rise of the Extinction Deniers." The author points out, "Just like climate deniers, they're out to obfuscate and debase the scientists and conservationists trying to save the world—and maybe get rid of a few pesky species in the process."[76] In May 2019, congressional Republicans used a hearing before the House Committee on Natural Resources to sow doubt on the extinction crisis, summoning well-known environmental and climate science deniers as their prime witnesses. Reporting on the hearing, a reporter for *The Guardian* wrote, "Republicans Aren't Just Climate Deniers. They Deny the Extinction Crisis, Too."[77]

In fact, conservative's extinction denial goes way back; conservatives have been challenging biodiversity loss and mangling the science of conservation biology for decades. By the early 1990s, biologists already were describing the loss of biodiversity as a global crisis. In 1992, the late E. O. Wilson, two-time Pulitzer Prize-winning Harvard biologist, warned about the human-driven "sixth mass extinction" in his book *The Diversity of Life*.[78] In 1993, the conservative Competitive Enterprise Institute published a pseudo-scientific rebuttal of the biodiversity crisis by Julian Simon, a well-known conservative economist, and political scientist Aaron Wildavsky. Following the conservative playbook, Simon and Wildavsky challenged the science on biodiversity in "Assessing the Empirical Basis of the 'Biodiversity Crisis'"; their work presaged the current denial of the extinction crisis:

> The scare about species extinction has been manufactured in complete contradiction to the scientific data. It is truth that is becoming extinct, not species. . . . At present, some conservation biologists seem more intent on whipping up concern for species loss than they are in documenting the extent of that loss and analyzing the possible ramifications, if any.[79]

A few years later, Paul and Anne Ehrlich wrote about conservatives' challenges to the extinction crisis, their distortions of conservation biologists' scientific findings, and their unrelenting efforts to reshape the Endangered Species Act in their 1996 book *Betrayal of Science and Reason*.[80]

Congressional Republicans have a long history of trying to repeal or gut the Endangered Species Act by spreading false claims about the scope of the problem and insisting on the use of "sound science" produced by industry scientists in congressional deliberations of the act. More recently, scientifically indefensible claims refuting the extinction crisis have appeared in the *Washington Times, The Daily Caller, The Daily Signal,* and on Fox News and other conservative media outlets. Breitbart called the 2019 report by IPBES "fake news" and claimed, "The two biggest human threats to wildlife in the last century have been (a) communists and (b) environmentalists."[81]

In 2020, the scientific journal *Nature* published an article titled, "Biodiversity Scientists Must Fight the Creeping Rise of Extinction Denial."[82] The authors chronicled the three-pronged approach to the denial of biodiversity loss. Literal denial involves claiming scientific facts are untrue. For example, the Heartland Institute posted an article that challenged the IPBES's 2019 report, falsely asserting that there has been a "significant and steep decline of extinctions since a peak in the late 19th and early 20th centuries."[83] Another form of extinction denial called interpretive denial involves cherry-picking data to make specious claims out of context such as using data from temperate ecosystems to refute biodiversity loss in the tropics where most losses are occurring. Lastly, the authors discuss implicatory denial in which data is not denied but rather nonsensical solutions to the extinction crisis are advanced. The central role of habitat destruction and climate change in the biodiversity crisis is well-established in the scientific literature. However, for conservative defenders of the dominant worldview, technology is the favored remedy for all environmental ills. Not surprisingly, one form of implicatory denial involves claims that Jurassic Park–like technological solutions can fix the extinction crisis.

Like the conservatives who glimpsed the threat posed by the alternative environmental worldview soon after the birth of the environmental movement, this new breed of extinction deniers also recognizes the political implications of the extinction crisis and the science on biodiversity loss. In his May 2019 statement before the House Committee on Natural Resources, Subcommittee on Water, Oceans and Wildlife, Patrick Moore, a policy advisor on climate and energy at the conservative Heartland Institute and self-published author of *Fake Invisible Catastrophes and Threats of Doom,* challenged the 2019 global assessment report by the IBPES.[84] Moore claimed that "highly exaggerated claims" by the IBPES are "a front for a radical political, social, and economic 'transformation' of our entire civilization."[85] In other words, just like other forms of environmental science, biodiversity science is a threat to defenders of the dominant worldview.

The science of the extinction crisis joins the tsunami of environmental evidence driving a potential societal paradigm shift.

ENVIRONMENTAL SCIENCE AND PARADIGM SHIFTS

Since the early days of the environmental movement, conservatives have challenged and distorted environmental science across a broad swath of issues including acid rain, ozone loss, deforestation, biodiversity loss, and climate change in tandem with discrediting environmentalists. All of these issues have serious consequences for human welfare. However, given the scope and stakes associated with global warming, climate change has become the epicenter of conservatives' campaign of distortion. According to most conventional analyses, by challenging the science of global warming and climate change, conservatives are protecting economic and political fortunes dependent upon fossil fuels and are warding off burdensome emission reduction regulations incompatible with conservative political ideology. However, we also need to factor in the role of worldviews.

Conservatives have taken the denial and distortion of climate change science to new levels in their quest to defend the dominant worldview. Conservatives are defending a worldview that assures them that the climate crisis has been greatly exaggerated, a worldview that fuels certainty about the reliability of human ingenuity and technological prowess. They are defending a worldview that assures them that infinite economic growth—powered by fossil fuels—is normal and possible in a world made for humans. Thus, for a richer understanding, we need to unearth the subterranean role of conflicting worldviews in environmental politics. In this long-running saga of denial and deceit, we need to recognize that conservatives have diligently and deviously undermined the momentum of the environmental movement and the authority of environmental science to block a societal paradigm shift.

We are long overdue in calling out the conservative defenders of the dominant worldview who have discredited the environmental movement with caricatures and unfounded accusations, and weaponized lies about environmental science to create a bulwark against change. If we do not openly confront the dangerous distortions of environmentalism and environmental science that prevent us from tackling existential environmental threats, gazing into our grandchildren's worried faces, what will we say?

FIVE

DECONSTRUCTING
ENVIRONMENTAL SCIENCE

One might argue that the siege on environmentalism and environmental science has been a signature accomplishment of the U.S. conservative movement. Conservatives have succeeded in delaying an effective national response to global warming for four decades. With the sole exception of the Clean Air Act of 1990, a market-based approach to mitigating acid rain, Congress has not enacted significant environmental legislation since the 1970s.[1] Conservatives' long-running campaign to caricaturize environmentalists has hobbled the movement's ability to catalyze social and political change. Some have contended that conservatives' decades-long campaign of climate science denial has contributed to the disregard for science more broadly in society. Naomi Oreskes has argued that climate change denial paved the way for COVID-19 denial in American society.[2]

How did the conservative assault on environmental science become so deeply embedded in our politics? Why do so many members of the public question the legitimacy of environmental science? Climate change communication researchers find that people refuse to believe information that does not align with their worldview. Conservatives' distortion and rejection of environmental science perfectly aligns with the dominant societal worldview: in a world made for humans, a world capable of absorbing all manner of human damage, clearly the environmental "crisis" has been exaggerated by environmental "extremists" and "alarmist" scientists. Other research suggests that members of the public believe what they are told by political and public opinion leaders. No doubt, rank-and-file Republicans have heard endless segments about fake climate science on Fox News, other conservative media outlets, and on Facebook, Twitter, Truth Social, and other social media platforms. However, conservatives also have been aided in their siege on environmental science by the emergence of a school of thought in the social sciences known as social constructionism. Oversimplified notions of social constructionism gave conservatives a weapon to undermine the legitimacy and authority of environmental science. This chapter provides an overview of social constructionism, its influence on environmental science, and its practice in social science scholarship on the Anthropocene.

AN INTRODUCTION TO SOCIAL CONSTRUCTIONISM

Social constructionism emerged in the intellectual context of postmodernism in the 1980s. It has become an important school of thought in the social sciences that has enriched our understanding of complex social phenomena. Social constructionism focuses on the social construction of reality and knowledge. Constructionists argue that the production of scientific knowledge is a social process, influenced by social and institutional norms and practices, individuals' identities, and methodological conventions and traditions within disciplines. Constructionists have made significant contributions to our understanding of gender, race, ethnicity, and sexuality as social constructs rather than biologically based categories. Social constructionists also have demystified the scientific process by studying how communities of scientists work together in the pursuit of scientific knowledge. Oreskes and other constructionist scholars emphasize that acknowledging and understanding the social production of scientific knowledge is key to the legitimacy of scientific findings.[3]

Constructionist critiques often focus on the identities of scientific practitioners. Feminist constructionist scholars have asked, "How could science claim to be objective when it largely excluded half the population from the ranks of its practitioners?"[4] Feminist scholars argue that the best way to develop objective knowledge is to increase the diversity of scientific communities. Scientific communities made up of diverse individuals representing a variety of viewpoints and identities are better equipped to collectively address multiple sources of potential bias, thus enhancing the advancement of objective, scientific knowledge. As an illustration of this principle in practice, Oreskes points out that the IPCC "makes a particular point of seeking geographical, national, racial, and gender diversity in chapter-writing teams."[5]

Some constructionists are quite clear about the distinction between social and physical reality. For example, political scientist Hoyoon Jung highlights the influential role of social constructionism in international relations in elucidating the ways in which political reality is socially constructed.[6] However, some social constructionists disregard the distinctions between social phenomena and physical reality. Many implicitly or explicitly discount the physical reality of the natural world. J. Gayon describes social constructionism as being "opposed to the thesis of the reality of an independent world."[7] Elizabeth Bird criticized the work of natural scientists, saying that "[t]he conduct of the natural sciences has proceeded under the assumption that scientific knowledge is a representation of something that exists outside it."[8] Although S. Barry Barnes argues that most social constructionists do not deny the existence of the external world, "strong" or "radical" social constructionists speak explicitly of the "social construction of natural reality itself."[9] Similarly, Sven Ove Hansson points out that constructionist scholarship is replete with statements suggesting that "the natural world is a social construction that does not exist independently of human thinking."[10]

Regardless of their views on the physical reality of the non-human world, constructionists maintain that natural scientists cannot objectively analyze an independent physical reality. Constructionists insist that scientists do not *discover* the nature of reality, they *construct* it. Many constructionists view the natural world as having a minor or even nonexistent role in the production of scientific knowledge.[11] Many constructionist critics of natural science reject the objectivist claims of science, instead viewing science as having " no unique or reliable fit to the world, no certain correspondence with reality."[12] Similarly, Hansson notes that for many constructionists, "science has no better claim than any other belief system to objective truth about nature."[13] Alan Sokal and Jean Bricmont point out that in many intellectual circles, it is assumed that all facts are socially constructed, and scientific theories are merely myths or narratives.[14]

For social constructionists, all knowledge is a social construction regardless of whether data is obtained by examining intangible social interactions or tangible, physical phenomena. For example, social scientists who study power in organizations cannot physically measure power with some type of power gauge. Using surveys and interviews, social scientists evaluate organizational members' perceptions of power, including members' perceptions of individuals who effectively wield power in the organization. The titular head of an organization may not be the actual leader if other organizational actors exert a controlling influence. In other words, power in organizations is an intangible social construct. In contrast, forest ecologists can measure tree diameter, height, and age in sample plots to calculate the amount of carbon stored in biomass. Biomass calculations inform forest ecologists' recommendations on management decisions aimed at maximizing carbon sequestration in forest stands. Unlike power, carbon—a key chemical ingredient in most forms of life on earth—is a physical, measurable substance.

It is true that most fundamentally, all knowledge is "constructed" by humans. Knowledge is not delivered to humans by the gods or some other source of external revelation. However, social constructionism can be taken to the point of anthropocentric absurdity. Constructionism disregards the fact that the world consists of more than humans and their perceptions. There *is* an external, physical world that exists outside of human consciousness. Should we assume humans are incapable of gleaning direct knowledge of the physical world even though our survival as a species always has depended on reading and understanding the environment? Sokol and Bricmont point out that the social constructionist literature avoids "any explicit acknowledgment of the role of the external world, and particularly of its possible causal impact upon human beings."[15] In the twenty-first century of the Anthropocene, how can we possibly ignore the external world of nature and its impact on human beings? Indisputably, the physical world is wreaking havoc on humans all over the world.

SOCIAL CONSTRUCTIONISM AND CLIMATE SCIENCE DENIAL

The early days of conservatives' calculated assault on environmental science occurred simultaneously with a growing focus on climate change science by social constructionists (also known as epistemic relativists) in the 1990s. According to Hansson, constructionists "treated climate science as a clear and therefore useful example of claims by natural scientists that should be seen as mere social constructions, rather than as reports reflecting the actual state of the natural world."[16] If social constructionism had remained within the ivory towers of academe, perhaps it would not have had an influence on environmental politics. However, constructionist critiques of global warming and climate change science found their way into the world of conservative think tanks, chief purveyors of climate denialism.

Hansson conducted a wide-reaching study of the constructionist literature to examine whether constructionists in academia who challenged the empirical bases of global warming had contributed to climate science denial more broadly. Hansson found that many prominent scholars engaged in constructionist critiques of climate science, such as American sociologist Frederick H. Buttel and his colleagues, British sociologist Nicholas Fox, environmental historian William Cronon, anthropologist Mary Douglas, political scientist Aaron Wildavsky, leading sociologist of science and the environment Brian Wynne, British sociologist Steven Yearley, and acclaimed sociologist of science Steve Fuller. For example, Buttel dismissed the scientific consensus on global warming and criticized other sociologists who had accepted global warming as a scientific fact. Buttel argued that the focus on climate change was "as much or more a matter of the social construction and politics of knowledge production as it is a straightforward reflection of biophysical reality."[17] Calling the environmental movement's affirmation of the reality of global warming a "noble lie," Buttel argued that global warming "provided a persuasive rhetorical framework for the organizational interests of the environmental movement."[18] Douglas and Wildavsky's famous cultural theory of risk was influential in leading other social scientists to characterize climate change as a cultural construction of risk rather than as a response to physical dangers independent of individual or cultural thought constructions.[19]

Hansson found ample evidence of constructionist scholars being influenced by science denialists' claims. For example, Buttel and Cronon used information from contrarians affiliated with the conservative Marshall Institute and the Cato Institute, respectively, to challenge climate science. However, Hansson also examined whether climate science denialists were inspired or influenced by academic constructionist critiques of climate science. Hansson found that Wildavsky had close connections with right-wing think tanks, "indicating intellectual community and exchange of ideas concerning climate change."[20] In 1992, Wildavsky wrote the introduction to Robert C. Balling's climate change denial book, *The Heated*

Debate: Greenhouse Predictions Versus Climate Reality.[21] Wildavsky became a board member of the Science and Environmental Policy Project founded by one of the original "merchants of doubt" and notorious climate denialist Fred Singer. Steve Fuller, a senior research fellow at the Breakthrough Institute, claimed that climate change deniers make a positive contribution to constructionists' intellectual claims by "showing the robustness of its core insights."[22] Hansson also found that some constructionist scholars have challenged the scientific consensus on other issues including evolution and AIDS in collaboration with conservative think tanks.

Myanna Lahsen notes that some constructionist scholars raised concerns about the political implications of their work.[23] Harry Collins et al. point out that constructionist scholarship had "exactly the scepticism about experts and other elites that now dominates political debate in the US and elsewhere."[24] Some scholars worried that assertions about climate change, as "just a construction by climate scientists seeking to increase research funding and/or promote their socio-political agenda," would arm conservatives with arguments to block a policy response.[25] Late in his career, prominent social constructionist Bruno Latour lamented the unintended cultural and political consequences of social constructionism, observing, "Entire Ph.D. programs are still running to make sure that good American kids are learning the hard way that facts are made up."[26]

Hans Radder posited an analogy to demonstrate the logical albeit ridiculous consequences of strong constructionist critiques of environmental science:

> Consider an environmental issue such as "the hole in the ozone layer." ... According to the ontological relativist point of view, this hole is identical to the discourse about it, and it cannot possibly have any independent reality. Consequently the hole would simply disappear at the very moment we stopped discoursing about it, even if ... we continued employing present technologies, such as aerosols, in an unaltered way![27]

According to this logic, if we simply stop engaging in discourse about global warming, the problem will cease to exist. Lahsen observed that concern about deconstructions of environmental science were evident in peer-reviewed literature but typically were confined to private conversations.[28] According to Hansson, with few exceptions, constructionist critiques of climate science disappeared after the 1990s, once constructionists' contrarian positions became associated with corporate and right-wing denialists.[29]

SOCIAL CONSTRUCTIONISM AND THE SCIENCE WARS

It was not long before social constructionism emerged from the intellectual bubble of academia and conservative think tanks. Social constructionism gained public prominence during the Science Wars of the 1990s. The Science Wars occurred simultaneously with the early days of conservatives' war on environmental science and the proliferation of constructionist critiques of climate science. The Science Wars consisted of a series of public debates between social constructionists and scientific realists. Social scientists from a variety of disciplines studying science as a social construction criticized the nature of scientific debate, the scientific worldview, and the institution of science. On the other side of the Science Wars were physical scientists defending empiricism, arguing for the legitimacy of the scientific method and the objective reality of scientific knowledge. Physical scientists strenuously objected to the constructionist notion that "Science is just a social justification system, with the implication being that theories are arbitrary and carry no more truth validity than other human narratives."[30] Biologist Paul Gross and mathematician Norman Levitt argued that postmodern constructionism was politically dangerous and jeopardized trust in science. Similarly, other physical scientists argued that social constructionism had affected perceptions of the scientific disciplines in both academic and popular culture in the United States and Europe. Bricmont argued that a major cultural problem fueled by postmodern social constructionism is "the destruction of empiricism . . . as an attitude toward what constitutes genuine knowledge."[31]

The Science Wars brought social constructionism into the cultural mainstream. Some commentators have disputed the significance of the Science Wars, arguing that they were waged by a small minority of physical scientists against a particular school of constructionist sociologists. However, articles about the Science Wars were published in major media sources including the *New York Times* and *The Atlantic*, giving social constructionism greater public prominence.[32] Oversimplified notions of social constructionism found their way into American society and have

shaped public attitudes toward science. In the eyes of the public, if scientific opinion is merely a social construction, the authority of empirical
environmental science is thrown into question. As Ava Kofman observed,
"If scientific knowledge was socially produced—and thus partial, fallible,
contingent—how could that not weaken its claims on reality?"[33] Characterizing scientific knowledge as a social construction also suggests that
science is provisional and, therefore, uncertain. As Bricmont pointed out,
"the emphasis is constantly put on the relativity of our knowledge, on the
importance of uncertainty."[34] Constructionist scholars' focus on scientific
uncertainty bolsters the claims of scientific uncertainty manufactured by
climate denialists.

The influence of social constructionism in environmental politics was
magnified by conservatives' weaponization of constructionist claims. Conservatives have challenged the science of global warming as "socially constructed and politically biased balderdash."[35] David Demeritt noted that
the political strategy of "social construction as refutation" was pursued by
climate skeptics opposed to the 1997 Kyoto Protocol.[36] Politicized notions
of social constructionism sever scientific opinion from the empirical foundations of the natural sciences. Unmoored from physical realities, scientific knowledge becomes merely one perspective in environmental policy
debates. Characterizing physical science as a socially produced perspective
by a community of like-minded scientists facilitates the politicization of
environmental science.

Science as a social construction can be interpreted to suggest that scientific opinion is crafted to meet a social group's political or ideological
agenda. Science construed as a social construction transforms the empir-
ically based *scientific* consensus on issues such as global warming and biodiversity loss—based on mountains of data and decades of peer-reviewed
studies—into a *political* consensus. Characterizing the scientific consensus
on global warming or biodiversity loss as a political consensus also reinforces parallel arguments about the environmental movement as a left-
wing cause espousing political arguments camouflaged as science. Later
in his career, Bruno Latour lamented that constructionists' criticisms of
science "had created a basis for antiscientific thinking and had paved the
way for the denial of climate change."[37]

The politicization of social constructionism facilitated the propagation
of other "perspectives" on environmental issues written by lawyers, policy
analysts, and economists based on ideologically driven arguments and

cherry-picked statistics completely at odds with actual, empirical environmental science. The American media's commitment to balanced news coverage has aggravated the problem by presenting "both sides," even though
one side is scientifically indefensible. For example, conservation biologists
warn that we are in the midst of a human-driven sixth mass extinction.
Climate change and biodiversity loss represent primary threats to the stability of the Earth System. In contrast, conservative economists, political
scientists, lawyers, and pundits with no training in biology claim there is
no extinction crisis. In 2019, the conservative newspaper the *Washington
Times* claimed species extinction is "highly exaggerated" and "authorita-
tive propaganda."[38] These specious "scientific" claims about biodiversity
loss, climate change, and other major planetary threats—put forth by
commentators with no training in the natural sciences—are repeated by
politicians and legitimized and amplified by journalists and media outlets as other perspectives on environmental issues in the name of balanced
coverage. These pseudo-scientific perspectives undermine the authority
of actual environmental science, suggest that scientific opinion is riddled
with uncertainty and contested, confuse the public, and undermine the
environmental movement's ongoing efforts to mobilize supporters. Some
might insist that politicians and pundits always have fabricated facts to
buttress their political positions and ideological preferences. However,
oversimplified notions of social constructionism legitimized baseless assertions that undermine the authority and significance of empirical environmental science.

DECONSTRUCTING THE ANTHROPOCENE

Conversations about the Anthropocene first emerged in the natural sciences. Not long after the Subcommission on Quaternary Stratigraphy established the Anthropocene Working Group in 2009, a growing chorus
of social scientists argued for greater inclusion of the social sciences and
humanities in Anthropocene research. For example, Eva Lövbrand et al.
claim, "By cultivating environmental research that opens up multiple interpretations of the Anthropocene, the social sciences can help to extend
the realm of the possible for environmental politics."[39] The global climate
system is central to the stability of the Earth System; thus, climate change
science is a major component of Anthropocene research. Political scientist
David Victor says that the IPCC needs to "overhaul how it engages with the

social sciences."[40] Victor demands greater inclusion of sociologists, political scientists, and anthropologists along with the economists already centrally involved in the IPCC's major assessments. He emphasizes that the social sciences are "central to understanding how people and societies comprehend and respond to environmental changes, and are pivotal in making effective policies to cut emissions and collaborate across the globe."[41]

Analyses of the Anthropocene have proliferated across a broad swath of disciplines including philosophy, history, sociology, political science, anthropology, and archeology. Articles on the Anthropocene have been published in a variety of disciplinary as well as transdisciplinary journals. Social scientists recognize that they have a crucial role to play in the overall Anthropocene conversation. Social science research can help us better comprehend the social, political, economic, and cultural drivers of the Anthropocene; their expertise can inform societal efforts to build a more just and sustainable future. Lövbrand et al. emphasize that critical social science scholarship can expose the assumptions and power relations embedded in the institutions charged with addressing climate change. They point out that "The fundamental challenges to societal organization posed by the Anthropocene are, paradoxically, to be countered by many of the same institutions that have allowed the recent human conquest of the natural world."[42]

However, some of the constructionist literature on the Anthropocene disregards the empirical realities of the Anthropocene and inadvertently contributes to the weaponization of constructionism as a tool to delegitimize environmental science. Numerous constructionists have characterized the Anthropocene as simply a social construction rather than a physical, planetary reality profoundly shaping human experience. For example, Lövbrand et al. characterize the Anthropocene as "a concept in the making."[43] The Anthropocene originally may have been a concept in the making; however, the Anthropocene Event is an empirical phenomenon. Others have contested the very idea of the Anthropocene, suggesting it is simply an academic invention or pop culture artifact. Claiming that the Anthropocene is simply a worldview, Jeremy Baskin writes, "the Anthropocene is less a scientific concept than the ideational underpinning for a particular worldview. It is paradigm dressed as epoch."[44] Is the Anthropocene simply an idea or conceptual framework, a social construction? Is the term merely jargon used by a particular community of scientists to characterize a way of thinking about current planetary conditions?

Social constructionists have objected to the dominance of the natural sciences "in crafting a narrative of the Anthropocene."[45] They characterize the physical science as simply one narrative among other legitimate interpretations of the Anthropocene that incorporate historical, cultural, and social analyses. For example, Lövbrand et al. argue that "the mainstream story projected by leading environmental scientists . . . has offered a restricted understanding of the entangled relations between natural, social and cultural worlds."[46] In response to the scientific narrative, numerous social scientists have argued that evocative names such as the Technocene, Econocene, Capitalocene, Westocene, Richocene, and Consumocene would better capture the multiple, complex social drivers of the Anthropocene. Unfortunately, characterizing the physical science narrative of the Anthropocene as simply one perspective undermines the authority and legitimacy of the empirical science that informs our understanding of the Anthropocene.

Rather than simply demanding the inclusion of social science research in the overall Anthropocene conversation, some constructionists argue that the scientific narrative of the Anthropocene should incorporate social science analyses as well. For example, Baskin asserts that Earth System scientists' interpretations of nature in the Anthropocene ignore the social sciences and humanities.[47] The social construction of nature is a dominant theme in the constructionist literature on the Anthropocene. Constructionists focus on peoples' perceptions of nature rather than the physical reality of the natural world. Peoples' perceptions and images of what constitutes "nature" are quite varied. For some people, an urban park or backyard garden is nature. To others, only vast tracts of wilderness "where the earth and its community of life are untrammeled by man" are what constitute nature.[48] In other words, *perceptions* of nature are social constructions.

However, environmental scientists study the physical realities of nature; they do not investigate peoples' perceptions of nature. The scientific narrative of the Anthropocene is grounded in the natural sciences, not the social sciences and the humanities. Nevertheless, Baskin argues for an expanded role for social scientists in evaluating scientific data, claiming, "By bringing in the social, one brings in . . . contestations of 'facts,' or at least of which 'facts' might count."[49] In what real, physical world does it make sense for social scientists—with no training in the natural sciences—to contest "the facts" that inform our understanding of physical nature in the Anthropocene? Should natural scientists contest "which facts might count" in social science research?

Taking his argument a step further, Baskin claims "the natural world cannot be studied independently of the social which helps to construct it."[50] In other words, Baskin is asserting that we cannot study the physical world of nature without simultaneously examining social systems and the complex political and economic factors reshaping nature and driving us into the Anthropocene. It is true that humans and nature—the natural and the social—are inextricably bound together in contrast to the dominant worldview's assumptions about human exemptionalism and the separation of humans and nature. However, the existence of social-ecological systems does not mean that natural scientists cannot examine the "natural" components of social-ecological systems without simultaneously studying human social systems and structures. Baskin's assertion suggests that we would need biologists and social scientists trooping around in all kinds of weather to monitor wildlife populations and species migrations as even plants move north, seeking cooler temperatures in a warming world. Is it impossible to measure atmospheric levels of CO_2, or nutrient flows in waterways, or the acidity of ocean water, or the trillions of tons of ice flowing into the ocean from the melting Greenland ice cap without the involvement of social scientists? Why would we expect natural scientists to incorporate in-depth social analyses in their scientific assessments of the physical world of the Anthropocene?

Arguing for the inclusion of social science perspectives in natural scientists' assessments of the physical world might make sense if social science scholarship regularly incorporated the role of biological, geological, or climatic processes in shaping human societies. However, there is little evidence that social scientists are receptive to reframing social science concepts based on Earth System science. As Nigel Clark and Yasmin Gunaratnam observe, "Social science concepts and theories appear able to be deployed as if they are immune to any reciprocal 'contamination' . . . we seem to be prohibited from inquiring about the geologic or climatic processes that might have shaped human collectivities or social formations."[51]

DECONSTRUCTING EARTH SYSTEM SCIENCE

Characterizing the Anthropocene as a social construction is problematic. However, constructionists' critiques go beyond dismissing the Anthropocene as simply an idea or conceptual framework. Some constructionists have challenged and deconstructed the Earth System science that constitutes the bedrock of our understanding of the trajectory of the Earth System in the Anthropocene Event. Manuel Arias-Maldonado asserts that "social scientists . . . should not contest scientific expertise as if they were natural scientists themselves."[52] Yet, some constructionists have challenged and deconstructed empirical natural science based on the conviction that scientific knowledge is simply a social construction and, therefore, subject to social science analysis. These intellectual arguments open the floodgates to analyses unmoored from empirical realities that can perpetuate the distortion of science in society and environmental politics. A distorted version of this constructionist position legitimizes pseudo-scientific, politicized claims made by lawyers, economists, and other social scientists about environmental problems as other reasonable perspectives on environmental policy issues.

For example, some constructionists have argued that systems thinking and the Earth System are social constructions "projected onto reality, rather than as an intrinsic property of the observed world."[53] In other words, they claim that these scientific concepts and frameworks—born out of decades of observation of the physical world and evidence-based research—do not describe empirical phenomena in the physical world. Instead, they insist the Earth System and systems thinking are social constructions created by and imposed upon the world by physical scientists with a global governance agenda. Based on this logic, should we also assume that tidal charts or phases of the moon replete with poetic language about waxing and waning are merely social constructions concocted by oceanographers and astronomers rather than depictions of observable, physical reality?

Earth System scientists employ a variety of research methodologies, measurement tools, and computer models to study the Earth System. Will Steffen's article, "The Emergence and Evolution of Earth System Science," provides an excellent summary of Earth System scientists' research methodologies.[54] According to Steffen et al., direct observations and experiments inform our understanding of the past and current Earth System, including the global climate, a key component of the overall Earth System. One of the best-known direct observations of the global climate system is the iconic Keeling Curve. Originally the brainchild of Charles Keeling, the Keeling Curve is based on daily measurements of atmospheric CO_2 at the Mauna Loa Observatory in Hawaii, beginning in 1958. The daily CO_2 measurements on Mauna Loa comprise the longest-running record of direct measurements of atmospheric CO_2.

Other direct observations of the global climate system include top-down observations obtained via remote sensing that provide data on important climatic variables including changes in land cover, atmospheric composition, the ocean surface, and urban development. Large-scale observations by interdisciplinary research teams help connect data from the local and planetary levels. NASA tracks how humans have changed the composition of the atmosphere, including CO_2 and ozone-depleting gases. Some studies use remote-sensing and ground-based approaches to study the dynamic interactions of the atmosphere, biosphere, and hydrosphere in the Amazon rainforest, a vast ecosystem essential to global climate stability.

Data collected about the Earth System's past elucidate present-day dynamics. Scientists reveal features of ancient climates by examining ocean salts, volcanic ash, soot from forest fires, coral reefs, sediments, tree rings and glacial ice. The Vostok Antarctica ice core, for example, reveals 420,000 years of paleoclimate data. Climate scientists trek for miles in the harshest climates on earth to drill ice cores; in Antarctica, some ice cores are two miles long. Frozen ice cores then have to be transported hundreds of miles—frozen and fully intact—to labs around the world where scientists analyze the atmospheric composition of gases and particles trapped in the ice. The ice also can disclose more recent phenomena. Atmospheric scientists at the University of Washington analyzing Greenland's ice found evidence of the passage of the Clean Air Act: they found clearer ice after 1970.[55]

Sophisticated mathematical models of the Earth System also are a key component of Earth System science. Climate models incorporate physical, chemical, and biological laws based on well-established theory (such as the laws of thermodynamics) and empirical data obtained through direct observations in order to simulate the entire climate system. The models simulate the interactions of the key drivers of the global climate system, namely the atmosphere (including air temperature, moisture and precipitation levels, and storms), the oceans (including measurements of ocean temperature, salinity levels, and circulation patterns), terrestrial processes (including carbon absorption, forests, and soil moisture storage), and the cryosphere (sea ice and land-based glaciers).

Although the models are not perfect representations of the global climate, physics-based climate models have enabled climate scientists to predict the pace and amount of global warming and some of the consequences—such as sea level rise, extreme precipitation, and more powerful hurricanes—decades before they could be physically observed. In 2021, the Nobel Prize in Physics was awarded to two scientists for their physical modeling of the global climate system.[56] The awardees' research was honored for laying the foundation for current climate modeling and for demonstrating the reliability of climate modeling. By what stretch of the imagination is this massive, transdisciplinary body of empirical and model-based data that constitutes Earth System science simply a social construction?

In Noel Castree's discussion of Anthropocene research in social science and humanities scholarship, he argues that critical social scientists can help "foster a mature debate about what grounds scientific statements about 'an earth in crisis.'"[57] Why would social scientists claim such a debate is needed? The theoretical foundations of Earth System science and a staggering amount of empirical data ground "scientific statements about an earth in crisis." But according to constructionists, a debate is needed because they view the Anthropocene as a social construction, subject to social science analysis. In other words, constructionists are unwilling to accept Earth System scientists' empirical and modeled findings as definitive; they are refusing to engage with the physical reality of the Anthropocene.

For example, Lövbrand et al. assert, "We believe that the social sciences can help to open up new possibilities for environmental debate by illustrating that there is nothing foundational in nature that needs, demands or requires sustaining."[58] Nothing that needs sustaining? Biologists report that the world is facing an unfolding "insect apocalypse" and a pollinator crisis. Many favorite foods like apples, strawberries, melons, and even chocolate cannot be produced without a variety of bees, bats, birds, butterflies, ants, flies, and wasps elegantly adapted to pollinate specific plant species. Perhaps we should sustain insects and pollinators? Perhaps we should sustain forests, mangroves, tidal marshes, seagrass meadows, and underwater kelp

forests that sequester tons of carbon, critical to the stability of our global
climate system? Presumably the millions of people whose lives have been
upended by global warming–fueled disasters, and the millions more who
fear continued climate chaos, want to live in societies with some predictable degree of climate stability.

Social science research informs educational practice, cultural norms,
politics, and public policy. As an influential school of thought, social constructionism has shaped scholarship and teaching pedagogy in many
disciplines. Constructionists have deconstructed politicized and often
discriminatory notions of contested social categories such as race, gender,
ethnicity, and sexuality. By shining a light into the black box of the scientific method, constructionists' research has provided important insights
about the legitimacy and authority of science of great relevance to environmental politics. Moreover, we need the expertise of social scientists to
help us build the social architecture of a more sustainable and just future
in the Anthropocene. Fruitful collaboration between social and natural
scientists is essential if we are to successfully address the challenges of the
Anthropocene.

However, oversimplified notions of social constructionism have perme-
ated public attitudes toward empirical science and have had an insidious
influence on environmental politics. Simplistic notions of constructionism gave conservative defenders of the dominant societal worldview an
ill-founded weapon to politicize environmental science, thus undermining the authority and significance of empirical environmental science. It
isn't simply students enrolled in PhD programs learning that "all facts are
made up."[59] For many members of the public, facts are simply opinions in
disguise, even scientific facts. Dueling opinions about global warming, biodiversity loss, and other planetary threats put forth by so-called experts
without any scientific expertise confuse the public, create cultural memes
of scientific uncertainty that undercut the well-established scientific consensus on our planetary problems, and undermine social momentum for
political change. In other words, social constructionism has had real-world
political consequences.

Do social constructionists have a responsibility to address the social
and political consequences of their work? Should they consider the political ramifications of characterizing science as a belief system and scientific
opinion as simply one viewpoint? By 2004, Latour had asked "self-critically
whether his own academic activities had contributed to [climate science

denial]."[60] In a world of planetary destabilization and irreversible ecological losses, do constructionists have an obligation to evaluate the implications of their claims about physical reality and environmental science? Manuel Arias-Maldonado and Zev Trachtenberg argue it is time for social scientists to acknowledge that "scientific results [cannot] be treated as merely one more way of representing reality that is no more valid than any other."[61]

If environmental problems were merely social constructions—if the problems were identical to our discourse—we would not have to worry about notions of planetary destabilization, runaway global warming, and the collapse of ecological systems. As Radder ironically suggested, if our discourse created the problem, we could just stop talking about them.[62] If only environmental problem-solving were this simple. In the absence of magical thinking, we need social scientists to ground their important scholarship in environmental realities that exist outside of human consciousness. It is time to focus on the empirical construction of physical reality.

SIX

DECONSTRUCTING
SCIENTIFIC UNCERTAINTY

Scientific uncertainty is unavoidable in environmental policymaking. However, the influence of social constructionism in environmental politics has elevated the role of scientific uncertainty and undermined the power of scientific consensus. Characterizing scientific knowledge as a social construction suggests that science is provisional and, therefore, uncertain. As Jean Bricmont noted, "the emphasis is constantly put on the relativity of our knowledge, on the importance of uncertainty."[1] Just as conservatives weaponized social constructionism to delegitimize climate science as "socially constructed and politically biased balderdash," they also have weaponized scientific uncertainty to undermine the scientific consensus, confuse the public, and block government action.[2] This chapter analyzes the role and implications of scientific uncertainty in environmental politics, policy, and pedagogy.

SCIENTIFIC UNCERTAINTY IN ENVIRONMENTAL POLICYMAKING

Judith Layzer and Sara Rinfret assert that most "policy-relevant questions require experts to make value judgments based on uncertain data."[3] Despite the robust scientific consensus on the reality and severity of our chief planetary threats, environmental policymakers still must grapple with varying degrees of scientific uncertainty, competing interpretations of the relevant scientific information, and difficult tradeoffs between environmental, economic, and technological feasibilities in any area of environmental policy. For example, although biologists have reams of data on threats to ecosystems at all spatial scales, on-the-ground management recommendations for specific threatened ecosystems involve some degree of scientific uncertainty compounded by questions of economic feasibility. Given the pervasive and ongoing nature of human threats to vital ecosystems, ecologists increasingly look to resilience theory to guide the restoration and protection of threatened ecosystems with important ecological, cultural, or economic values. However, the ecological resilience of particular ecosystems is still a relatively new area of biological inquiry. As Anne Chung et al. emphasize, "Although resilience-based management is now well established theoretically, there have been few examples of implementation."[4]

Climate science is unequivocable on the need to drastically mitigate greenhouse gas emissions; however, formulating mitigation policies is never a straightforward endeavor even if we adhere to the IPCC's findings and set aside conservatives' opposition to reducing fossil fuel emissions. Similarly, climate change affects so many dimensions of life—including public health, workplace safety, infrastructure stability, and transportation, among others—that policymakers working on climate change adaptation also must deal with scientific and economic uncertainties and technological complexity often paired with political controversy. For example, adaptation specialists must contend with uncertainties regarding the specific effects of global warming-fueled severe weather. In coastal areas, policymakers must deal with questions about the anticipated force and geographic extent of storm surge and flooding. Policymakers working on public health regulations regularly face scientific uncertainties caused by the compound nature of toxic pollution. People typically are exposed to so many different pollutants simultaneously that isolating the health impacts of any individual pollutant can be difficult. Establishing safe standards for individual pollutants is further complicated by gender, age, technological and political feasibility, and the ethics of human subject research.

However, in some cases, what is commonly called scientific uncertainty is actually a question of political or technological feasibility. For example, during the Obama administration, the EPA was trying to establish new regulations to control a major public health threat known as fine particulate pollution ("soot"). According to Walter Rosenbaum's account of the controversy,

> As the debate evolved, it became clear that setting the new standard also involved substantial scientific uncertainties. EPA scientists had long acknowledged that setting the particulate standard was difficult because any standard except total elimination of the airborne particulates would still create significant public health risks. In short, no scientifically "safe" standard could be created, and any decision would have to be made on the basis of how much risk to public health would be considered acceptable.[5]

In this example there was no actual scientific uncertainty from a public health perspective. There was a scientific consensus that no safe level of airborne particulates could be established. Determining the acceptable level of risk to public health is the thorny political dilemma. The scientists may have disagreed on the health effects of specific levels of particulate pollution recommended by policymakers; however, the scientists did not disagree on the dangers of particulate pollution: the safe level is zero.

The foregoing suggests that the nature of scientific uncertainty in environmental politics demands a nuanced treatment. General references to scientific uncertainty distort important political dynamics. The fossil fuel industry and their conservative allies have weaponized the uncertainty inherent in the scientific method to undermine the authority of climate science. Denialists also have fabricated uncertainty to confuse the public and dampen public support for governmental action. Manufactured claims of scientific uncertainty create a recipe for political inaction; climate science is characterized as too provisional to warrant a policy response. Thus, critical analyses of scientific uncertainty in environmental politics require distinguishing between the uncertainty inherent in the scientific method, manufactured scientific uncertainty that warps environmental politics, and legitimate, unavoidable scientific uncertainty that accompanies specific environmental policy initiatives. In other words, we need to differentiate between the realities and implications of scientific uncertainty in environmental *policymaking* versus the role and impact of scientific uncertainty in environmental *politics*.

SCIENTIFIC UNCERTAINTY IN GLOBAL WARMING SCIENCE

Scientific uncertainty is inherent in the scientific method. The integrity of scientific findings rests in large part on transparent appraisals of probabilities and uncertainty. In 2010, the IPCC released metric-based guidance for the consistent treatment of scientific uncertainty in all groups working on major assessment reports.[6] Although there is no scientific uncertainty about the reality of global warming, in such a complex area of Earth System science, uncertainty about dynamic interactions among positive and negative feedbacks that accelerate or counteract global warming is unavoidable. For example, climate scientists are still evaluating the effect of clouds on global warming. Some models show clouds cooling the atmosphere; other studies show clouds magnifying global warming. Areas of scientific uncertainty shrink and shift as scientists deepen their understanding of the global climate system. As climate science evolves, study methodologies are modified to address different questions, new data is analyzed, and scientists report novel or revised findings.

For example, for years it was well established in the scientific literature that the Arctic is warming twice as fast as the rest of the planet, a phenomenon known as arctic amplification. This is an issue of significant concern because the Arctic plays a major role in global weather patterns. However, in 2021, NASA researchers reevaluated the time period and the latitudes typically employed in Arctic warming studies. By examining warming during the past thirty years, Peter Jacobs et al. found that the Arctic is actually warming four times faster than the rest of the planet.[7] Previous studies using longer time frames did not account for the acceleration of global warming. Moreover, earlier studies had examined Arctic warming above the 60° N parallel, a boundary that included much of Scandinavia. Jacobs et al. shifted the study area slightly farther north to 60.66° N, the traditional demarcation of the Arctic Circle. They found warming above 60.66° N was four times faster than the rest of the globe. The fact that global warming is accelerating means that researchers must constantly adapt; climate change data unavoidably will change. As Jacobs et al. explained, "When something is changing as quickly as the climate, numbers can get old and outdated quickly."[8] However, the fact that data change does not mean that scientists' predictions are faulty and riddled with scientific uncertainty.

THE FAT TAIL OF CLIMATE RISK

When confronted with scientific uncertainty, what conclusions should we draw? Denialists typically assume scientific uncertainty means that environmental problems are less severe than claimed or predicted. This assumption is consistent with the conviction embedded in the dominant social paradigm that environmental problems have been greatly exaggerated. Conservatives' weaponized claims about scientific uncertainty often are paired with accusations about alarmist scientists and environmental extremists who exaggerate the threats of environmental problems. Denialists claim that scientists exaggerate the risks by only reporting data that support the worst-case scenarios.

For example, a 2019 IPCC report found that global mean sea levels will most likely rise between 0.95 feet and 3.61 feet by the end of this century.[9] Denialists argue that scientists and environmentalists always focus on the worst-case scenario: over three feet of sea level rise. In reality, climate scientists working in teams that base their findings on scientific consensus tend to underestimate the dangers of climate change.[10] However, when climate data change, like the data on Arctic warming or sea level rise, denialists blame "alarmist" scientists. In an interview with the Heartland Institute, contrarian scientist Patrick Michaels claims that changes in global warming data "always seem to point in the direction of 'it's worse than we thought.'"[11]

The fact that climate change data keep changing is not simply due to scientific uncertainty or alarmist scientists or environmental extremists. The data keep shifting because global warming is accelerating faster than originally predicted. It is getting worse. It is dangerous to assume that global warming and many other environmental problems will be less severe than predicted. Scientific uncertainty cuts both ways. The recent research on Arctic warming indicates that global warming in the Arctic is twice as bad as previously reported. As UN Secretary-General António Guterres warned, "The disruption to our climate and our planet is already worse than we thought, and it is moving faster than predicted."[12]

In *Climate Shock: The Economic Consequences of a Hotter Planet,* Gernot Wagner and Martin Weitzman introduce the "fat tail of climate risk"[13] (Figure 6.1). The fat tail of risk means that "the likelihood of very large impacts is greater than we would expect under typical statistical assumptions."[14] Under the fat tail, there is a 10 percent chance of extreme impacts;

FIGURE 6.1. THE FAT TAIL OF CLIMATE RISK: AN ESTIMATE OF
THE LIKELIHOOD OF WARMING DUE TO A DOUBLING OF PRE-
INDUSTRIAL GREENHOUSE GAS CONCENTRATIONS.

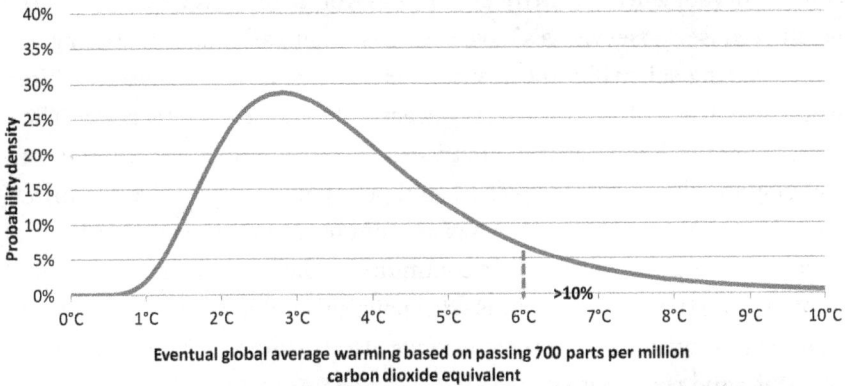

Eventual global average warming based on passing 700 parts per million
carbon dioxide equivalent

Source: Adapted from Gernot Wagner and Martin L. Weitzman, *Climate Shock: The Economic Consequences of a Hotter Planet,* 2016, p. 53. Copyright © 2016 by the Princeton University Press. Reprinted with permission.

under a normal bell curve, there is only a 2 percent chance of outliers. In other words, we need to confront assumptions embedded in the dominant social paradigm; we should not treat scientific uncertainty in climate change politics as our friend. We should not assume that global warming will be more benign than predicted, and we should not assume that there is only a tiny chance of severe impacts of global warming. The likelihood of severe impacts is greater than would be expected under a normal bell curve of distributed risks. As Mark Cliffe emphasized, the "sting of climate risk is in the tails."[15] Contrary to denialists' claims, scientific uncertainty cuts both ways.

SCIENTIFIC CONTROVERSY

In environmental politics and policy, scientific uncertainty and scientific controversy go hand in hand. As Rosenbaum notes, "What often distinguishes environmental policymaking from other policy domains is the extraordinary importance of science, and scientific controversy, in the policy process."[16] Where there is genuine scientific controversy, the rational observer concludes that scientific uncertainty precludes the formation of a scientific consensus on the nature and risks of the problem. In other words, scientific controversy implies that the issue in question involves legitimate scientific uncertainties. However, scientific controversy in environmental politics often is not due to legitimate uncertainties within the scientific community. Scientific controversy in environmental politics often is the norm because conservatives have been systematically manufacturing and manipulating scientific uncertainty for more than three decades. Often what we call scientific controversy is actually political controversy. Thus, a more nuanced and accurate analysis of scientific uncertainty and controversy in environmental politics and policy must emphasize the differences between *political* controversy *over* environmental science and actual controversy *within* the scientific community. The fact that environmental politics is virtually synonymous with political controversy does not mean that every environmental issue we face also is riddled with scientific uncertainty and controversy.

Textbook author Zachary Smith offers a striking example of the emphasis on scientific controversy in environmental politics in his brief account of the publication of Rachel Carson's landmark book, *Silent Spring*. Smith writes that the publication of *Silent Spring* in 1962, "gave much of the attentive public a view of disagreement and controversy within the scientific community."[17] However, the story of *Silent Spring* is a chronicle of political controversy over science rather than a story of scientific disagreement within the scientific community.

Rachel Carson was a highly respected biologist and gifted writer. Four and a half years in the making, Carson's *Silent Spring* summarized an extensive body of research produced by biologists around the world.[18] In an accessible prose style, she alerted the public to the dangers of DDT and other similar persistent chemicals in the environment. Her work eventually led to important policy changes: DDT was banned in the United States in 1972. *Silent Spring* captured the attention of the American public (it was on the *New York Times* bestseller list for thirty-one months) and provided an impetus for the birth of the environmental movement. There was extensive media coverage of *Silent Spring* and much controversy. However, a lot of the published criticism of *Silent Spring* consisted of personal, sexist attacks on Rachel Carson. A reviewer in *Time* magazine criticized Carson for her "emotion-fanning words" and characterized her argument as "unfair, one-sided, and hysterically overemphatic."[19] In a derogatory review of *Silent Spring*, a spokesman for the American Medical Association wrote, "*Silent Spring*, which I read word for word with some trauma, kept reminding me of trying to win an argument with a woman. It cannot be done."[20]

Carson's work also was viciously attacked by representatives of the chemical industries that feared a ban on DDT, and by some agricultural industries concerned about a complete ban on pesticides (an argument that Carson never made). The Velsicol Chemical company threatened to sue Houghton Mifflin, the publisher of *Silent Spring*. The media lapped up Monsanto's "The Desolate Year," a parody of *Silent Spring* that completely distorted Carson's argument and scientific evidence.[21] Objections by male-dominated chemical industries opposed to regulation do not constitute controversy within the actual scientific community.

Generalized references to scientific uncertainty and controversy also undermine the significance of the scientific consensus surrounding global warming, ocean acidification, biodiversity loss, stratospheric ozone depletion, and other environmental threats of the Anthropocene. Although a robust scientific consensus does not translate to straightforward policymaking, the scientific consensus has important political implications. Public understanding of the scientific consensus on our major planetary threats can raise public concern, strengthen public support for a policy response, and serve as a catalyst for political mobilization. Downplaying the scientific consensus on major environmental issues erodes public worries and plays into the hands of the conservative opponents of environmental protection who have manipulated the contours of environmental politics.

Conservative defenders of the dominant worldview have contested the scientific consensus on major planetary issues for decades. To undermine the scientific consensus on anthropogenic global warming, conservatives have maintained a steady drumbeat of accusations of scientific uncertainty that has convinced 80 percent of Americans that numerous climate scientists disagree that global warming is human caused.[22] In reality, John Cook et al. found that 100 percent of climate scientists agree that global warming is human caused.[23] Conservatives have long recognized the political power of the scientific consensus. As the political consultant Frank Luntz emphasized, "Voters believe that there is no consensus about global warming within the scientific community. Should the public come to believe that the scientific issues are settled, their views about global warming will change accordingly."[24] Undermining the scientific consensus fuels cultural memes of scientific uncertainty reinforced by constructionist framing of science and conservatives' decades-long campaign against environmental science. Thus, it is incumbent on analysts to provide a nuanced treatment of scientific uncertainty and consensus in environmental politics.

PEDAGOGICAL DILEMMAS

Providing a nuanced treatment of scientific uncertainty in environmental politics can be challenging for environmental policy instructors and textbook authors who typically are constrained by academic and political norms of neutrality. However, conservatives' war on environmentalism and environmental science has warped the meaning of true political neutrality. In order to tell the actual story of environmental politics, instructors must be willing to debunk the false equivalencies that plague environmental politics. How can instructors tell the actual story of climate change politics without differentiating between conservatives' strategically manufactured scientific uncertainty and the genuine scientific uncertainty inherent in climate science? Similarly, how can instructors tell the actual story of environmental politics without differentiating between the scientific uncertainty that accompanies environmental regulatory and legislative policymaking initiatives and the weaponized scientific uncertainty that distorts environmental politics? Would sidestepping traditional academic and political norms in order to make these important distinctions—telling the actual story of environmental politics—pose risks to instructors' academic credibility and job security?

Unfortunately, the answer to the above question could be yes. In 2023, the Ohio General Assembly debated the Ohio Higher Education Enhancement Act (SB 83).[25] Ohio's bill was perhaps the most extreme of similar state-based initiatives aimed at diversity, equity, and inclusion initiatives in higher education modeled after the American Legislative Exchange Council's Intellectual Diversity in Higher Education Act.[26] SB 83 would have required instructors at state-funded colleges and universities to provide balanced coverage of a lengthy list of "controversial" political issues including "climate policies, electoral politics, foreign policy, diversity, equity, and inclusion programs, immigration policy, marriage, or abortion."[27] Cloaked as a measure to foster intellectual freedom and critical thinking, the bill prohibited the inculcation of "any social, political, or religious point of view" in higher education.[28]

The original version of the bill used the term "climate change" rather than "climate policy." However, since conservatives have defined climate change as a contested political issue, the change in terminology makes no substantive difference. It is impossible to discuss climate change policies without acknowledging the scientific reality and dangers of climate change. Thus, instructors who teach climate science or climate change politics and policy would have been required to frame climate change as a controversial political issue; they would have been forced to balance actual climate change science with denialists' proven falsehoods.[29] SB 83 would have mandated the posting of syllabi on publicly available websites, post-tenure reviews, and possible termination for tenured faculty who violate the policies outlined in the act. In other words, SB 83 would have established potential, professional consequences for instructors who failed to conform to the bill's requirements.

A modified, veiled version of the bill, "Regards State Higher Ed Institution Commitment to Certain Beliefs" (HB 394), was introduced in the Ohio House in February 2024. HB 394 would "prohibit state institutions of higher education from requiring individuals to commit to specific beliefs, affiliations, ideals, or principles."[30] But what does this language mean? If an A grade on an exam or paper requires an accurate discussion of climate change science and impacts, could that be construed as requiring students to commit to "specific beliefs, affiliations, ideals, or principles"? Could instructors be fired for requiring scientific accuracy? Political scientists, textbook authors, and instructors must tread carefully.

ENVIRONMENTAL POLITICS, SCIENCE, AND DEMOCRACY

Characterizing environmental science as a socially constructed perspective capitalizes on longstanding democratic norms. We view politics as contests among stakeholders with diverse and equally valid viewpoints. However, environmental issues challenge the normative bedrock of American politics. When it comes to issues grounded in science, not all perspectives are equally valid. Politically motivated, manufactured pseudo-scientific claims simply are not defensible when compared to evidence-based environmental science that aligns with ecological and planetary realities. The flood of pseudo-scientific claims that warp environmental politics and possibilities suggests that reconciling science and democracy is becoming more urgent. How do we reconcile science-based decision-making and democratic norms in a policy domain that involves irreversible, ecological consequences that contravene norms of policy reversibility and compromise intergenerational equity?

The science-based nature of environmental problems has challenged commentators and political theorists for decades. Since the birth of the modern environmental movement, some have argued that democracy is ill-equipped to adequately address environmental problems. Over the years, various authors have argued that some form of expert-based authoritarian government would do a better job of solving environmental problems. Beginning in the 1970s, "eco-authoritarian" arguments appeared in the writings of Robert Heilbroner and William Ophuls.[31] These early writings were solidly dismissed. Nevertheless, a new generation of eco-authoritarians have argued that "governments should be granted full discretion to carry out public programs and to intervene in the personal and economic activities of citizens, without having to abide by limitations emerging from citizens' private and democratic rights."[32]

More recently, a number of scholars and climate scientists have grappled with the shortcomings of democracy in the context of addressing climate change. According to Mayson Glenn Obrien, proponents of environmental authoritarianism argue that,

> Properly responding to the overwhelming magnitude and intensity of the current climate crisis will require a similarly overwhelming policy response . . . environmental authoritarianism offers centralized government control over policy decisions allowing the government to streamline policy implementation.[33]

Dale Jamieson, professor emeritus of environmental studies at NYU, has argued that we do not have political institutions capable of addressing climate change, the largest collective action problem we have ever faced.[34] The late British scientist James Lovelock likened climate change to war and argued that we need to abandon democracy to rise to the challenges of climate change.[35]

Others have argued that democracy is best suited to effectively addressing climate change. Defending the central importance of democracy to civilization, Nico Stehr argues that

> The alternative to the abolition of democratic governance is more democracy—making not only democracy and solutions more complex, but also enhancing the worldwide empowerment and knowledgeability of individuals, groups, and movements who work on environmental issues.[36]

Similarly, Daniel Fiorino makes a compelling case for democracy in *Can Democracy Handle Climate Change?* Arguing that the critics of democracy "do not understand politics," Fiorino maintains that taken together, the dynamism and innovation potential of the private sector, the capacity for policy learning, democratic societies' commitment to gender equity, state-level innovation and leadership, and democratic nations' active global engagement together well-equip democracies in the struggle to combat climate change.[37] Fiorino also emphasizes that we owe future generations systems of democratic governance as well as environmental sustainability.

There is a related literature on democracy and sustainable development relevant to environmental politics given the centrality of environmental sustainability in sustainable development discourse. Andrea Westall examines the tensions between democracy and sustainable development;[38] Dan Banik argues that democracy is essential for the achievement of sustainable development.[39] In the global context of Anthropocene governance, a host of scholars—among them Frank Bierman, John Dryzek, Eva Lövbrand, Gerard Delanty, and Amanda Machin—have grappled with how to reconcile democratic decision-making within nation-states with the scientific expertise necessary for managing the Earth System in the Anthropocene.[40]

Beyond the macro-level challenges to democratic governance posed by complex science-based environmental issues such as climate change, science-based policymaking poses challenges for government officials who do not possess the necessary expertise, as well as for citizens and stakeholders who expect their concerns to be considered in policy decisions. Well-designed processes must facilitate democratic participation, policy learning, and effective use of relevant scientific information. In other words, as David Pedersen notes, "scientific expertise needs to be mediated through a complex process of social and political deliberation."[41] Over the years, many scholars and environmental practitioners have studied and tested a variety of approaches to better integrate democratic decision-making and science-based policymaking. Their recommendations enhance science-based decision-making in democratic deliberation by creating processes that enable stakeholders to challenge scientifically spurious claims. Recommendations include creating boundary-spanning structures and processes to provide policymakers with essential scientific information, participatory processes to build trust and facilitate collaborative decision-making, and adaptable institutions. The growing literature on collaborative governance reflects these trends.[42]

Niklas Wagner et al. assert that recurring or ad hoc science-policy interfaces (SPIs) are essential for supporting decision-makers with relevant scientific findings.[43] They emphasize that the social learning that occurs in SPIs reveals and clarifies problem complexities and helps bridge the divide "between 'experts' and 'non-experts' through information exchange."[44] They also point out that SPIs can facilitate "the interaction between science and policy as an iterative, joint co-production process" and "improve the communication of complexity and uncertainty," both of which are particularly important in the context of environmental sustainability issues.[45] Focusing on the needs of citizen stakeholders, Fiorino has examined ways to improve participatory mechanisms including public hearings, initiatives, public surveys, citizen review panels, and collaborative processes in the regulatory arena known as regulatory negotiations (or negotiated rulemaking) to enhance meaningful citizen participation in technology and science-based policymaking.[46] Alice Brites et al. analyzed science-based stakeholder dialogues as a strategy for bringing science closer to decision-making in the design of environmental policies.[47] Similarly, Herman Karl et al. have examined policy dialogues as a means of integrating science and policy through joint fact-finding.[48]

Environmental dispute resolution scholars and practitioners also have pioneered processes that protect both scientific integrity and democratic deliberation. Disputes over data are legendary in environmental disputes and policymaking. As Rosenbaum points out, "data become weapons" in environmental politics.[49] Environmental mediators have introduced innovative processes collectively referred to as data negotiation for dealing with disputes over environmental science in site-specific environmental disputes and policy dialogues. Rather than waging what dispute resolution scholar Lawrence Susskind called the "battle of the printout," stakeholders engage in joint fact-finding and agree on the sources of scientific information used to inform their deliberations.[50] Stakeholders occasionally design needed studies together, agreeing on the geographic scope, time frame, and key variables to be examined. Political actors trying to advance specious scientific claims have to credibly justify their arguments. Pseudo-scientific positions that are inconsistent with the scientific evidence agreed upon by the negotiators are rejected.

The possibility of runaway global warming and biodiversity loss that could destabilize planetary systems suggests that we should re-evaluate how we treat scientific uncertainty in environmental politics and pedagogy. Younger generations coming of age in the "world of wounds" of the

Anthropocene will not thank us if we are enablers of misguided skepticism about environmental science in education and politics. We need to provide a nuanced treatment of scientific uncertainty to elucidate the important distinctions between scientific consensus and legitimate versus politically manufactured scientific uncertainty. Arguably, the stability of the Earth System and the sustainability of our democratic system of governance ultimately depend on renouncing contrived, pseudo-scientific arguments about scientific uncertainty that have stymied environmental policymaking for decades. It is time to reclaim the legitimacy and authority of environmental science in environmental politics and pedagogy.

SEVEN

DODGING DENIAL

The Politics of Geoengineering

The dominant societal worldview and free-market economy fiercely defended by conservatives is based on the unrestrained use of cheap fossil fuels. From 1850–1950, coal dominated the U.S. energy landscape. After World War II, petroleum consumption took off, overtaking coal in 1950.[1] After 1950, petroleum use and greenhouse gas emissions both soared exponentially. The IPCC repeatedly has emphasized that preventing dangerous global warming demands ending our dependence on fossil fuels. In June 2023, UN Secretary General António Guterres issued a clarion call to world governments, urging "Countries must phase out coal and other fossil fuels to avert climate 'catastrophe.'"[2] Yet, for three decades, conservatives have challenged and dismissed the IPCC's warnings and downplayed the dangers of global warming. Simultaneously with their climate change denial, conservatives have supported research on technologies to manage the global climate known as geoengineering.

Geoengineering fits squarely within the dominant societal worldview. Geoengineering is grounded in the dominant worldview's unwavering faith in human technological prowess and control of nature. For conservatives, geoengineering offers a means of perpetuating an economy driven by fossil fuels. After years of manipulating climate science, lying about the scientific consensus, manufacturing scientific uncertainty, and disparaging environmental and climate activists and scientists, conservatives' support for geoengineering takes on new dimensions. This chapter provides an overview of geoengineering and traces the history and implications of conservatives' support for geoengineering technologies. The chapter highlights the stakes associated with conservatives' long-running war on environmentalism and environmental science.

OVERVIEW OF GEOENGINEERING

Geoengineering refers to a suite of engineering-based technologies for managing the global climate. There are two general categories of geoengineering technologies: carbon dioxide removal (CDR), often called carbon capture, and solar radiation management, often called solar geoengineering. For many, the term "geoengineering" is synonymous with solar radiation management. Carbon capture is a two-step process that involves "capturing" atmospheric carbon dioxide and locking it away for centuries in soils, plants, rocks, saline aquifers, and exhausted oil wells so that it cannot escape into the atmosphere. Carbon capture includes a variety of nature-based approaches such as reforestation, planting trees in previously unforested areas (known as afforestation), and soil restoration to enhance the removal of atmospheric carbon dioxide through natural, biological processes in the soil. Other forms of carbon capture combine natural processes with technology such as capturing the carbon produced by burning biomass, referred to as bioenergy with carbon capture and storage. Carbon capture serves the dual purpose of removing some of the excess CO_2 already in the atmosphere and preventing more from entering the atmosphere.

Solar radiation management, often called solar geoengineering, typically is envisioned on a broad, even global scale. It involves a variety of tactics designed to reflect solar radiation away from earth before the energy can be trapped by CO_2 and the other greenhouse gases that produce global warming. A widely discussed form of solar geoengineering involves injecting sulfate aerosols into the stratosphere to reflect sunlight back into space, simulating what happens after a major volcanic eruption. For example, when Mount Pinatubo erupted in 2001, it sent roughly 15 million tons of sulfur dioxide into the stratosphere, causing a detectible cooling of the earth's surface for about two years. Solar geoengineering is more controversial and riskier than most forms of carbon capture.

Atmospheric chemist Paul Crutzen is often credited—or blamed—for spurring interest in geoengineering. In 2002, Crutzen wrote that sustainable environmental management in the Anthropocene "may well involve internationally accepted, large-scale geoengineering projects to optimize climate."[3] A few years later, Crutzen published an influential article in the prestigious journal *Climatic Change* in which he broke the taboo of discussing atmospheric geoengineering research within scientific circles. Unfortunately, Crutzen's initial comments on geoengineering frequently have been taken out of context.

Crutzen was concerned about the health impacts of atmospheric aerosols such as sulfur dioxide (SO_2), a form of air pollution that causes respiratory illnesses and other human health impacts. Ironically, SO_2 aerosols and other forms of air pollution also help to cool the planet by reflecting light, thus masking some global warming (much like naturally occurring volcanic eruptions). Crutzen recognized that the dual role of aerosols as a public health threat and as a global cooling mechanism posed challenges for policymakers seeking to address global warming. Crutzen was worried that policymakers would not adequately control aerosol pollution because of global warming concerns.

Most commentators on Crutzen's key role in launching serious discussion of geoengineering have ignored the context of Crutzen's remarks as well as the fact that Crutzen did not see solar geoengineering as a desirable solution. In time, Crutzen wondered if breaking the taboo on discussing geoengineering was a "moral imperative or moral hazard."[4] Given the potentially catastrophic effects of unchecked global warming—a logical fear in the absence of meaningful global emission reduction treaties and commitments at the time—Crutzen initially believed that exploring options to protect the global climate and humanity was a moral imperative. However, Crutzen ultimately concluded that reducing greenhouse gas emissions was the only viable path to addressing global warming. He worried that unleashing public discussion and research on geoengineering would legitimize an approach for reducing global warming that would be a disincentive for reducing greenhouse gases and be potentially dangerous if deployed prematurely, thus becoming a moral hazard. In his earlier works, Crutzen repeatedly emphasized that "By far the preferred way to resolve the policy makers' dilemma is to lower the emissions of the greenhouse gases."[5] In an interview near the end of his life, Crutzen emphasized that research on geoengineering should be viewed only as a last resort to prevent runaway global warming should policy options fail. Crutzen once again affirmed, "Reducing CO_2 emissions is the solution."[6]

To this day, questions abound about whether geoengineering is an indispensable tool to stabilize the global climate—a moral imperative—or a dangerous option with many unforeseen consequences—a moral hazard. Much of the support for geoengineering is centered on carbon capture; much of the concern is focused on solar geoengineering. Most forms of carbon capture can be deployed on local scales, with many forms relying on natural biological processes. In contrast, solar geoengineering typically is envisioned on a global scale and represents a more intrusive intervention in the global climate system.

Geoengineering as Moral Imperative

Some see geoengineering research and development as a moral imperative so that technologies are available—only if needed—to prevent catastrophic global warming. Most proponents of geoengineering insist that these technologies are intended only as a supplement, not a replacement for mitigation measures to reduce greenhouse gas emissions. The IPCC has examined the role of carbon capture in stabilizing the global climate for many years. In 2021, the IPCC noted that carbon capture may be necessary to reduce global temperatures in case the 1.5 degrees Celsius threshold established by the Paris Climate Agreement is breached. The IPCC also has noted that ongoing but limited commercial uses of fossil fuels are indispensable; thus, some forms of carbon capture will be necessary to counteract "hard-to-abate" emissions produced by agriculture, aviation, shipping, and some industrial processes.[7] The IPCC highlights the important role of nature-based approaches to carbon capture, such as tree planting, combined with technological controls to drawdown atmospheric carbon dioxide. Among climate scientists willing to consider geoengineering, their ambivalent acceptance is based on their twin worries about the increasingly severe impacts of climate change now and unchecked future global warming. Weighing the dangers of runaway global warming versus geoengineering, some climate scientists ask, "Compared to what? More dangerous than the climate-change-driven famine, flooding, fires, extinctions, and migration that we're already beginning to see?"[8]

There also is growing interest in solar geoengineering, even though it is more controversial. In 2021, the National Academy of Sciences released a report stating the "US Should Cautiously Pursue Solar Geoengineering."[9] There are several major national and international research programs and assessments of solar geoengineering underway, and there are proponents within the scientific community. In 2023, the preeminent climate scientist James Hansen argued that global warming is proceeding faster than predicted by climate models; thus, solar geoengineering will be necessary to cool the planet before the effect of emission reductions can restore the planetary energy balance.[10]

Some analysts have challenged the dominant framing of large-scale geoengineering; they argue that small-scale deployment of solar geoengineering can and should be deployed before we are confronted with more severe, irreversible impacts of global warming.[11] Others within the scientific community have reluctantly sanctioned research programs on solar geoengineering as an emergency brake if policy responses fail. In the context of climate justice, some have argued that developing countries should lead solar geoengineering research.[12] Most developing nations have contributed very little to global warming, yet many in the global south are among the most vulnerable to climate change impacts.

Geoengineering as Moral Hazard

The dominant societal worldview with its unshakable trust in technology underscores the moral hazard of geoengineering. Carbon capture and solar geoengineering could disincentivize the necessary transition away from fossil fuels. Numerous climate scientists opposed to geoengineering also ask, what would be the consequences of a large-scale experiment? Climate scientist Michael Mann points out, "It is simply impossible to know or game out all the unintended consequences of deploying an untested technology on such a massive scale."[13] A more subtle dimension of the moral hazard lies with the tempting nature of new technologies. Some analysts have pointed out there are few technologies that have been developed that have not eventually been deployed. Many worry that research on solar geoengineering in particular will inevitably lead to real-world experiments.

On multiple occasions, the IPCC has warned about the moral hazard and substantial risks of solar geoengineering. One obvious problem with deploying solar geoengineering without significant emission reductions would be ongoing ocean acidification—the less familiar, dangerous consequence of global warming. The IPCC has warned that solar geoengineering could introduce a "widespread range of new risks to people and ecosystems, which are not well understood."[14] Notably, the IPCC does not include solar geoengineering in their assessments of various emission reduction pathways. A consequence of solar geoengineering more relatable for the average person involves the color of the sky. If we embark on solar geoengineering by scattering sunlight, "the sky will never be as blue again."[15] Pulitzer prize-winning author Elizabeth Kolbert named her 2021 book featuring geoengineering *Under a White Sky*.[16]

Opponents also ask, what would happen if we stopped engineering the climate? The IPCC has warned, "a sudden and sustained cessation of solar geoengineering would drive a rapid increase in global temperature within a decade or two, endangering biodiversity, weakening carbon sinks, increasing precipitation and changing water cycles."[17] Most scientists agree that if we stopped the solar geoengineering experiment, a dangerous increase in global warming would occur too swiftly for effective adaptation by people and other species. The implications of stopping large-scale solar geoengineering without concurrent large-scale emission reductions suggest that humans would need to continue solar geoengineering for millennia.

To assess experts' views on solar geoengineering, Astrid Dannenberg and Sonja Zitzelsberger surveyed 723 experts from more than 150 countries involved in scientific and diplomatic efforts to address climate change through the IPCC and the United Nations Framework Convention on Climate Change (UNFCCC).[18] Their study showed that respondents from countries facing potentially severe climate change impacts were more likely to support geoengineering. However, they also found that respondents from the IPCC were more opposed to solar geoengineering than respondents from the UNFCCC; the IPCC is widely recognized as the most credible source of scientific information about climate change. It is noteworthy that they also found that the natural scientists (the majority of the respondents in their sample) were more likely to oppose solar geoengineering research, deployment, and inclusion in international climate negotiations than respondents with professional backgrounds in economics, business administration, engineering, political science, or law.[19] Interpreting their findings, they suggested that individuals engaged in climate change diplomacy may be more open to geoengineering than scientists who focus on the physical impacts of climate change. It is striking that the natural scientists who study the global climate system are the most strongly opposed to solar geoengineering.

In addition to the many environmental issues and problems, there are many questions about the governance of solar geoengineering research, experimentation, and implementation. Which global institutions would make decisions, and would their policies be viewed as legitimate? What would happen if a rogue nation launched solar geoengineering unilaterally? Would solar geoengineering exacerbate or instigate geopolitical conflict? At the fourth session of the United Nations Environment Assembly in 2019, Switzerland introduced a modest resolution on a preliminary governance framework for carbon capture and solar geoengineering that revealed international tensions.[20] The United States and Saudi Arabia were opposed to the Swiss proposal. In 2022, the IPCC noted that there still is no "dedicated, formal international solar geoengineering governance for research, development, demonstration, or deployment."[21] Sikinah Jinnah and Simon Nicholson argue that

> the greatest danger of climate engineering may be how little is known about where countries stand on these potentially planet-altering technologies. Who is moving forward? Who is funding research? And who is being left out of the conversation?[22]

In response to these concerns, using balloons and satellites, the United States has launched an early warning system to detect solar geoengineering efforts by other countries or rogue actors.[23]

In 2022, an international multi-disciplinary group of more than sixty climate scientists and governance scholars signed an open letter calling for an international "non-use agreement" on the research and deployment of solar geoengineering. The letter states,

> Solar geoengineering deployment at planetary scale cannot be fairly and effectively governed in the current system of international institutions. It also poses unacceptable risks if ever implemented as part of future climate policy.[24]

The signatories called for no public funding, no outdoor experiments, no patents, no deployment, and no support in international institutions for solar radiation management. Jennie Stephens and Kevin Surprise argue that solar geoengineering would "further concentrate contemporary forms of political and economic power," pointing out that solar geoengineering research is being advocated by "a small group of primarily white men at elite institutions in the Global North." They maintain that it is "unethical and unjust" to advance solar geoengineering research.[25]

Given the substantial risks of solar geoengineering, carbon capture receives more attention in policy circles. However, there are numerous problems to overcome. Many see carbon capture as a "dangerous distraction" that will divert attention from the necessary investments and policies to significantly reduce emissions.[26] Kevin Anderson et al. argue that widespread assumptions about the feasibility of planetary-scale carbon capture have derailed the Paris Climate Agreement's commitment to keep warming below 1.5 degrees Celsius.[27] Many analysts maintain that carbon capture at the necessary scale to drawdown adequate amounts of atmospheric carbon dioxide is speculative at best. The IPCC notes that few reliable models for reaching the 1.5 degrees Celsius target incorporate carbon capture technologies. In April 2022, the IPCC warned that implementation of carbon capture currently faces

> technological, economic, institutional, ecological-environmental, and sociocultural barriers. Currently, global rates of carbon capture deployment are far below those in modelled emission pathways for limiting global warming to 1.5°C or 2°C.[28]

In addition, the lack of markets for the captured carbon disincentivizes investments in carbon capture technologies. Technologically, carbon capture facilities are plagued with a rampant record of failure. The Government Accountability Office reports that the U.S. Department of Energy has spent over $1.1 billion to advance carbon capture demonstration projects that have mostly failed.[29] The GAO acknowledged that most projects had minimal hopes of success. Over time, the trajectory of failure may change with new public investments in carbon capture contained in the Inflation Reduction Act.[30]

On top of the financial and technological problems, carbon capture increasingly is bumping into political opposition. Currently there are about 5,000 miles of carbon dioxide pipelines stretching across the United States. Energy experts predict that 100,000 miles of these pipelines "could crisscross the country" as the climate crisis worsens in coming decades.[31] Opposition to carbon dioxide pipelines has emerged in the midwest as well as in Louisiana, a state already home to a huge number of petrochemical facilities and refineries. In June 2022, the New Orleans City Council passed a resolution banning the development of carbon capture facilities and pipelines.[32] In midwestern states like Iowa and North Dakota, an unlikely alliance of Native Americans, white farmers, and conservative private property rights advocates has been fighting the construction of several pipelines designed to carry captured carbon dioxide 2,000 miles across Iowa, Minnesota, Nebraska, and North and South Dakota. In 2023, Illinois landowners were able to derail a carbon dioxide pipeline that would have stretched across South Dakota, Nebraska, Minnesota, Iowa, and Illinois for 1,300 miles.[33]

Local residents oppose such pipelines for a variety of reasons. Farmers do not want to lose productive land to pipeline easements; private property rights advocates oppose the use of eminent domain in pipeline siting. However, many residents are concerned about public safety. Local residents and safety experts alike worry that the existing regulatory framework for pipeline safety is inadequate for the types of carbon dioxide pipelines that will need to be built. In 2020, a carbon dioxide pipeline in Mississippi ruptured, releasing 31,405 barrels of compressed liquid CO_2. The compressed liquid immediately converted to CO_2 gas and settled over the nearby town of Satartia. Residents felt the effects of the carbon dioxide plume almost immediately. According to *Inside Climate News,*

> People struggled to breathe. One person had a seizure. Some residents had fallen unconscious in the middle of the road, while others stood around in a daze, unresponsive to commands. Cars stopped running, since combustion engines require oxygen.[34]

Forty-five people were hospitalized and two hundred residents were evacuated. Three years later, some Satartia residents report lingering health effects from the incident, including "increased frequency and severity of asthma attacks, headaches, muscle tremors and difficulty concentrating."[35]

THE POLITICS OF GEOENGINEERING

Both the Republican and Democratic Parties have supported geoengineering research and development; however, their stated motives are quite different. Republicans have advocated geoengineering as a technological fix to perpetuate reliance on fossil fuels. In contrast, Democrats have supported geoengineering as a supplement to aggressive emission reductions. The Democratic Party has not advanced geoengineering while simultaneously denying the scientific consensus on global warming, spreading specious claims about climate science, and arguing against the need for climate change legislation and regulation. In 2019, Democrats convened the House

Select Committee on the Climate Crisis. As the name suggests, the committee's report, *Solving the Climate Crisis*, is aligned with the IPCC's recommendations. It addresses the need for transformative change in energy use and production, manufacturing, supply chains, infrastructure, transportation, land use, housing, and urban design in order to move away from fossil fuels. The report does not advocate geoengineering in lieu of emissions reductions and system-wide decarbonization. Energy analyst David Roberts described the plan as "the most detailed and well-thought-out plan for addressing climate change that has ever been a part of US politics."[36]

Perversely, many conservative defenders of the dominant worldview advocate geoengineering as an excuse for climate change inaction as well as a necessity due to the urgency of the climate crisis. Denying the reality of global warming has become more difficult in the face of the many vivid and increasingly deadly manifestations of climate change. But instead of a *mea culpa* chorus about their misinformation, or a reluctant embrace of policies to reduce fossil fuel use and emissions, in a Houdini-like escape from their entrenched denial conservatives have embraced geoengineering to manage our climate problems, often without ever acknowledging the reality and dangers of global warming. Reporting on a House Science Committee hearing on geoengineering in 2017, Brian Kahn wrote, "None of the Republicans could bring themselves to acknowledge that carbon dioxide is the root cause of climate change. Nor could they bring up that reducing carbon emissions is a way more proven and cost-effective avenue to address climate change."[37] For example, in an impressive feat of rhetorical gymnastics, Representative Lamar Smith, the former head of the Science Committee who has railed against "carbon dioxide hysteria" for years, said "Generally we know that the technologies associated with geoengineering could have positive effects on the Earth's atmosphere. These innovations could help reduce global temperature or pull excess greenhouse gases out of the atmosphere."[38]

Although a growing number of individual Republican politicians now acknowledge the reality of anthropogenic global warming, there are no proposals offered by the Republican Party to reduce greenhouse gas emissions by transitioning away from fossil fuels. Climate change plans released by House Republicans call for the expansion of domestic fossil fuel production and infrastructure including pipelines and liquid natural gas terminals, small modular nuclear reactors, hydrogen, and carbon capture. Although some of their plans also include expanding wind and solar power, notably their plans do not call for emission reductions. In 2021, the *E&E Daily* reported, "Senate Republicans want to fight climate change by burning more fossil fuels."[39] Clearly House Republicans also want to combat climate change by burning more fossil fuels.

Geoengineering offers a technological fix to reduce greenhouse gas emissions that perpetuates reliance on fossil fuels and avoids fundamental changes in the energy industry. Conservatives and their industry allies argue that coal, oil, and gas are essential to the healthy functioning of the economy and an important source of jobs. They maintain that geoengineering offers a more economically feasible approach to protecting the global climate than reducing carbon emissions. Geoengineering is consistent with the free-market economy prized by conservatives. As Michael Mann observed, "For the free-market fundamentalist, geoengineering is a logical way out because it reflects an extension of faith that the free market and technological innovation can solve any problem we create, without the need for regulation."[40]

However, more fundamentally, conservatives' support for geoengineering is embedded in their adherence to the dominant societal worldview. The dominant worldview is grounded in unshakable confidence in humans' ability to control nature. Faith in our technological prowess undergirds the certainty that advanced technologies such as geoengineering will ensure that the earth remains hospitable for humans. For many DSP proponents, "Technological innovation incentivized by capitalism and the free market . . . means that we can continue our energy-intensive, consumer-intensive, globalized ways of life . . . indefinitely."[41]

The Ebb and Flow of Conservatives' Interest

The history of conservatives' interest in geoengineering closely tracks climate change politics in the United States. Whenever there was a credible threat of federal regulations that would mandate emission reductions to facilitate the transition away from fossil fuels, conservatives' interest in geoengineering went up. In 1988, NASA scientist James Hansen testified before Congress that anthropogenic global warming was real and already occurring. Not long after Hansen's testimony, conservative supporters of the fossil fuel industry began exploring solar geoengineering and carbon capture simultaneously with their campaign to distort and delegitimize climate science. The Electric Power Research Institute (EPRI) funded early research on solar geoengineering in the 1990s. In 1992, EPRI founder

Chauncey Starr co-authored a paper with the well-known climate denialist Fred Singer, arguing that "The scientific base for a greenhouse warming is too uncertain to justify drastic action at this time. Should climate
risks ever prove significant . . . we could geoengineer our way out of the
problem."[42] During the same time period, Exxon scientists were actively
writing about solar geoengineering. Ironically, the fact that sulfur dioxide
aerosols produced by the combustion of fossil fuels provide some measure
of global cooling has been cited as proof that solar geoengineering works to
control global warming.

  Jean-Daniel Collomb's analysis of ten conservative think tanks that
promoted geoengineering illustrates that conservatives' interest in geoengineering was sparked during the George W. Bush administration.[43] The
Bush administration's opposition to emission reduction policies created a
favorable climate for geoengineering proposals. However, conservatives'
interest in geoengineering during the Bush administration also coincided
with a pending Supreme Court decision, *Massachusetts v. EPA*.[44] Massachusetts and eleven other states, the Center for Biological Diversity, and
multiple environmental advocacy organizations petitioned the EPA re-
questing that the agency regulate carbon dioxide under the Clean Air Act.[45]
Consistent with the Bush administration's position on global warming, in
2003 the EPA rejected the petition, arguing that the act did not give them
the authority to regulate carbon dioxide. However, in 2007, the Supreme
Court ruled that the EPA had the authority to regulate CO_2 as a form of
air pollution under the 1970 Clean Air Act.[46] The significance of this ruling
cannot be overstated. The Supreme Court's decision made it legal for the
EPA to regulate carbon emissions. The specter of future federal CO_2 regulations loomed large.

  During this same time, a variety of conservative think tanks, including
well-known purveyors of climate change denial, wrote reports and opinion pieces and were actively involved in workshops and conferences to
further geoengineering research and advance geoengineering policy proposals. For example, the American Enterprise Institute (AEI), well-known
for their climate change denial, had an active program focused on geoengineering. In 2006, AEI hosted an event called Strategic Options for Bush
Administration Climate Policy. The description of the event warned, "the
new Democratic majority in Congress is certain to make climate policy a
priority, and federal action on climate change in the near future is increasingly likely."[47] In his report after the AEI event, Lee Lane, climate denialist

and co-director of the AEI geoengineering project, advocated for increased research funding for geoengineering technologies that "would avoid harmful climate change while allowing emissions."[48] In 2007, the Heartland Institute, well-known supplier of climate change disinformation, published a newsletter titled "Geoengineering Seen as a Practical, Cost-Effective Global Warming Strategy."[49] The Competitive Enterprise Institute, the Hoover Institution, and the Manhattan Institute also released reports on geoengineering.

In the wake of the 2007 Supreme Court decision in *Massachusetts v. EPA*, the impending election of Barack Obama further stoked conservatives' fears of a serious policy response to global warming. During his first presidential campaign, Obama vowed to enact a cap-and-trade system designed to dramatically reduce carbon dioxide emissions by 2050. Shortly after the election, President-Elect Obama emphasized that "he had no intention of softening or delaying his aggressive targets for reducing emissions that cause the warming of the planet."[50] Collomb found a significant bump in interest in geoengineering during the Obama administration. Prominent climate change denialists well-placed in conservative think tanks funded by the fossil fuel industry promoted geoengineering throughout the Obama administration.

From 2008–2010, AEI was a strong advocate of geoengineering research. In 2008, AEI sponsored an event called, "Geoengineering: A Revolutionary Approach to Climate Change." The website description of the event repeated their earlier warning that "Congress is likely to enact federal climate legislation in 2009."[51] As the co-director of AEI's geoengineering project, Lane testified before the House Committee on Science and Technology in 2009 in support of geoengineering research. In his statement, Lane urged the Committee to view the hearing as the first step in a "serious, sustained, and systematic effort by the U.S. government to conduct research and development on solar radiation management"; Lane argued that solar geoengineering research was necessary because "for many nations, a steep decline in greenhouse gas (GHG) emissions may well cost more than the perceived value of its benefits."[52] Perversely, Lane also justified research on solar geoengineering by raising the specter of "extremely harmful climate change," claiming that solar geoengineering might be the best option for preventing worst-case scenarios in the event of runaway global warming.

In 2009, Congress began deliberations on President Obama's cap-and-trade bill, the American Clean Energy and Security Act.[53] AEI sponsored an event on "Governing Geoengineering" simultaneously with congressional deliberations on the bill. When the Obama administration's cap-and-trade bill failed, the administration turned to the regulatory arena. By 2009, the legal foundation for regulating carbon emissions under the Clean Air Act was well established. However, when the Supreme Court ruled in 2007 that the EPA had the authority to regulate carbon dioxide, they also instructed the EPA to determine whether greenhouse gases are a threat to human health and welfare. In 2009, the EPA issued the consequential "endangerment finding," ruling that carbon dioxide and five other greenhouse gases emitted from smokestacks and other man-made sources are a threat to human health and welfare and, therefore, subject to regulation.[54] The endangerment finding cemented the legal basis for regulating greenhouse gases.

In 2009, the Manhattan Institute published a report called "Climate Change: Another Option,"[55] critiquing the Obama administrations' proposed Clean Power Plan, the regulatory initiative for reducing emissions from the electric power sector. The Manhattan Institute's report argued that geoengineering could slow or reverse global warming at a significantly lower cost without major reductions in carbon emissions: "Successful geoengineering would permit earth's population to make far smaller reductions in carbon use and still slow or reverse global warming, but at a vastly lower cost."[56] Although AEI discontinued their geoengineering program after 2010, Collomb's analysis confirms that conservative think tanks, many of which were actively involved in climate change denial, published reports favorable to geoengineering approaches from 2006-2017.[57] In 2015, the EPA published the final rule for Obama's Clean Power Plan.[58] A winter 2015-2016 report published by the Cato Institute as part of their series on regulation emphasized the benefits of solar geoengineering saying, "We could achieve Mount Pinatubo-like results, and in a controlled way, by pumping comparable amounts of sulfur dioxide into the air." Pointing out that the cost of such an initiative would be "pennies per ton," they concluded, "So why not go the sulfur route?"[59]

In 2016, the Supreme Court stayed the implementation of the Obama administration's Clean Power Plan in an unusual ruling before the regulations were actually implemented. The Supreme Court's decision on Obama's carbon regulations combined with the election of Donald Trump eliminated conservatives' worries about federal regulations that would compromise use of fossil fuels. The Trump administration withdrew the

United States from the Paris Climate Agreement and launched an unprecedented initiative to eviscerate the EPA's environmental regulatory programs. Although the first Trump administration did not aggressively pursue geoengineering strategies, numerous high-ranking officials within the administration were longtime advocates of geoengineering. For example, Rex Tillerson, former CEO of ExxonMobil, and Trump's first secretary of state, is known for having described climate change as "an engineering problem, and it has engineering solutions."[60] David Schnare, one of the key players on Trump's transition team, had long been a proponent of federal support for geoengineering. Although conservative groups' advocacy of geoengineering diminished during the first Trump administration, Collomb reports that geoengineering "was raised 15 times by five different think tanks" from 2017-2018.[61]

The Role of the Fossil Fuel Industry

The fossil fuel industry has been a key player in the geoengineering debate, and a major funder of geoengineering research and their conservative supporters' election campaigns. The Center for International Environmental Law argues that the fossil fuel industry's influence in the overall solar geoengineering debate is less pervasive now than it was in the 1990s and 2000s up through the end of the Obama administration. However, they caution that it would be "impossible and unwise to ignore the recurring influence of fossil fuel industries and interests in the research and policy agenda for geoengineering."[62] The fossil fuel industry has a long history of researching, promoting, and patenting geoengineering technologies including carbon capture and solar geoengineering to protect the continued extraction of fossil fuels and their profits. Not surprisingly, ExxonMobil reports that it has been researching carbon capture for more than thirty years.

Much of the fossil fuel industry's research has focused on carbon capture because it facilitates "enhanced oil recovery," a technique for extracting new oil from depleted wells. Pressurized CO_2 obtained via carbon capture can be injected into depleted wells to force the hard-to-reach remaining oil to the surface where it can be commercially extracted. Carbon capture also has been touted as a means of saving the coal industry, a "clean coal" technology to permit the continued extraction of coal for decades to come in defiance of market trends and global warming realities. Proponents estimate that carbon capture could "spur consumption of 40% more coal and up to 923 million additional barrels of oil in the US alone by 2040."[63] Undaunted by the record of failure for carbon capture projects, the major oil companies include carbon capture in their current forecasts to justify continued expansion of fossil fuels in the context of a low carbon future.

Fossil fuel companies also have become major supporters of climate and energy research at some of the most elite universities in the United States. Benjamin Franta and Geoffrey Supran found that fossil fuel interests and investors "have colonized nearly every nook and cranny of energy and climate policy research in American universities."[64] For example, MIT's Energy Initiative is funded largely by Shell, ExxonMobil, and Chevron; MIT also has accepted $185 million from oil billionaire David Koch, a well-known financier of climate change denial. The founding director of Stanford's Global Climate and Energy Project is a petroleum engineer; the project is funded by ExxonMobil and Schlumberger. UC Berkeley's Energy Biosciences Institute received $500 million from BP. The majority of Harvard's energy and climate policy research also is funded by Shell, Chevron, BP, and other oil and gas companies. However, funding from fossil fuel companies is not necessarily funneled into geoengineering research at these elite institutions. For example, Harvard's Solar Geoengineering Research Program states that they do not accept donations from fossil fuel companies, or from foundations and individuals whose wealth comes from fossil fuel investments:

> We are concerned that fossil fuel companies or other interests will seek to exploit solar geoengineering as a pretext for delaying reductions in greenhouse gas emissions. We do not want donors who are (or could reasonably be construed as being) motivated to support solar geoengineering research to protect fossil fuel industries.[65]

Nevertheless, it would be naïve to dismiss the industry's overall influence on climate change, energy, and geoengineering research in academia. In 2022, more than 500 academics released a letter calling on universities in the United States and the United Kingdom to stop accepting funding from fossil fuel interests. According to Zack Budryk, "To be clear, our concern is not with the integrity of individual academics. Rather, it is with the systemic issue posed by the context in which academics must work, one where fossil fuel industry funding can taint critical climate-related research."[66] In an interview with *The Guardian*, signatory Michael Mann emphasized that fossil fuel companies that advertise their support of university climate and energy research are purchasing the credibility, authority, and objectivity of elite institutions "while funding research that often translates into advocacy for false solutions and 'kick the can down the road' prescriptions like massive carbon capture, which is unproven at scale, and geoengineering, which is downright dangerous."[67]

THE COSTS OF A WORLDVIEW POWERED BY FOSSIL FUELS

The inescapable realities of climate change provide damning evidence against the continued use of fossil fuels. However, the manufactured debate over human-caused global warming has obscured the broader implications of an economy powered by the unrestrained use of fossil fuels. In 2023, scientists released an update on the state of the Earth System.[68] Their findings are sobering: in addition to biodiversity loss and deforestation, our reliance on fossil fuels, petroleum-based agricultural chemicals including fertilizers and pesticides, and plastics derived from petrochemicals are destabilizing the Earth System.[69] Successful geoengineering technologies could cool the earth or bring down atmospheric levels of carbon dioxide to mitigate global warming. However, with or without geoengineering, an economy powered by fossil fuels demands continued drilling and mining and the expansion of fossil fuel infrastructure with many accompanying human health and geopolitical problems as well as irreversible ecological losses. An economy run on fossil fuels also means the continued expansion of the plastics industry, the toxic underbelly of fossil fuels.

Drilling and mining for fossil fuels destroy treasured landscapes and valued ecosystems all over the world. Decapitated with millions of tons of dynamite to extract coal, hundreds of mountains in the Appalachian range have been obliterated along with several thousand miles of headwaters, streams, and forest ecosystems. Left with the legacies of mining, human communities tucked into the ancient mountain hollows suffer the health effects of air pollution and heavy metal contamination in drinking water. Fracking has destroyed beloved rural landscapes and caused health problems for nearby residents. In northwestern Canada, the extraction of tar sands oil, one of the dirtiest fossil fuels on earth, has devastated indigenous communities and turned millions of acres of ancient boreal forests into desecrated, oily moonscapes dotted with huge toxic tailings ponds visible from outer space.

Offshore oil drilling causes massive oil spills like the 2010 Deepwater Horizon accident in the Gulf of Mexico. Described as an "underwater volcano" of oil, the Deepwater Horizon spill polluted marine landscapes and organisms, harmed tourism, and debilitated the livelihoods of commercial anglers for months. Throughout the major cities of the United States, countless numbers of people, including the poor and people of color, are suffering from asthma and other health effects caused by airborne fossil fuel emissions from vehicles and manufacturing. Pipelines transporting fossil fuels snake for miles across private lands and important ecosystems, endangering communities in their path, and threatening to pollute agricultural lands and sacred bodies of water. Protests have erupted all across the country as a kaleidoscope of conservative farmers and ranchers, military veterans, millennials and younger, and Native Americans try to block the construction of even more fossil fuel infrastructure.

Although the United States produces more crude oil than any other nation, we are still subject to the whims of a global energy market, vividly illustrated by the war in Ukraine.[70] Russia's 2022 invasion of Ukraine caused chaos in global energy markets and soaring oil and gas prices worldwide. In her book *Blowout*, Rachel Maddow chronicled the dangers we all face by relying on fossil fuels produced by increasingly rogue "petro-states."[71] Continued dependence on fossil fuels by wealthy nations like the United States also has ethical consequences in the context of sustainable development. Continued reliance on fossil fuels by the world's wealthiest nations could constrain development opportunities in other nations. Researchers have emphasized that wealthy nations must end oil and gas production within the next decade to give developing nations more time to generate sources of income and energy other than fossil fuel production. And then, of course, there is global warming, fueled by the combustion of fossil fuels. The staggering environmental, social, and economic costs of climate change are incalculable.

The implications of conservatives' advocacy of geoengineering to maintain an economy powered by fossil fuels comes into sharp relief in the context of the global "carbon budget." The carbon budget refers to the cumulative amount of carbon the world still can burn and stay within IPCC global temperature targets. Bill McKibben first brought public attention to the carbon budget in a 2012 article published in *Rolling Stone* titled "Global Warming's Terrifying New Math."[72] The article laid out stark statistics about the amount of carbon contained in the fossil fuel industry's proven reserves based on the major oil companies' own internal data. The data showed that there is five times more carbon in the fossil fuel companies' proven reserves than we can safely burn. In 2016, McKibben published an update to his 2012 article called "Recalculating the Climate Math." Based on a study obtained from the Norwegian energy consultants Rystad, McKibben reported that the coal mines and oil and gas wells currently in operation worldwide contain nearly twice as much carbon—942 billion tons of carbon—than the 500 billion tons that we can safely burn to stay under 1.5 degrees Celsius of global warming. McKibben concluded, "The new study shows, we can't dig any new coal mines, drill any new fields, build any more pipelines. Not a single one. We're done expanding the fossil fuel frontier."[73] Echoing McKibben's conclusions, the IPCC has warned that building more fossil fuel infrastructure will "lock-in" greenhouse gas emissions. In April 2022, during the press conference accompanying the release of the IPCC's Sixth Assessment Report, UN Secretary-General António Guterres, said "Investing in new fossil fuel infrastructure is moral and economic madness."[74]

Conservatives' embrace of geoengineering, paired with their refusal to engage in bipartisan discussions about how to address global warming through emission reductions, demonstrates that their campaign against environmentalism and environmental science goes much deeper than an aversion to regulation. There are a variety of non-regulatory policy tools and market-based incentives to drawdown carbon emissions that are compatible with conservative ideology. In other words, we could imagine a story of climate change politics in which both parties accept the reality of global warming. Presumably, congressional Democrats would advocate for aggressive regulations, and Republicans would lobby for market-based approaches. Republicans likely would push for carbon taxes, emission trading systems, and tax credits to incentivize the manufacture and purchase of low-carbon energy sources and energy efficient appliances—like the incentives and tax breaks built into the 2022 Inflation Reduction Act.[75] Both parties might also throw discussion of geoengineering into the mix of policy tools under consideration. Given political and electoral incentives, we also could imagine that this policy debate would be difficult and protracted, frustrating scientists and environmentalists who understand the need for prompt, decisive action.

Decades ago, when acid rain became a public issue, rather than denying the reality of acid rain, the George H. W. Bush administration championed a market-based, cap-and-trade system that significantly reduced SO_2 emissions without banning the use of high-sulfur coal. Why is this not the story of climate change politics? Why do present-day conservatives reject market-based, non-regulatory incentives to mitigate greenhouse gases? Why have conservatives refused to embrace the successful model of the Regional Greenhouse Gas Initiative (RGGI)? RGGI is a cooperative market-based, cap-and-trade regulatory program to reduce greenhouse gas emissions within the power sector in Connecticut, Delaware, Maine, Maryland, Massachusetts, New Hampshire, New Jersey, New York, Pennsylvania, Rhode Island, and Vermont.[76] Revenues generated by RGGI's cap-and-trade program have provided substantial benefits to local businesses and consumers including investments in energy conservation and bill assistance for low income households throughout the eleven-state region. Since the program's inception in 2005, RGGI's investments up through 2021 have resulted in billions of dollars in energy savings and millions of tons of avoided carbon dioxide emissions.[77] Clearly, the story of conservatives' political intransigence in climate change politics goes much deeper than their aversion to regulation. Likewise, the conservative war on the environmental movement and environmental science goes deeper than simply neutering a social movement that supports environmental protection through regulation.

Conservatives' entrenched position of denial is grounded in the clash of worldviews that drives environmental politics. Conservatives' vision of the American way of life is inextricably intertwined with fossil fuels. Justified by the myths of the dominant societal worldview, their vision of the good life is only made possible by infinite economic growth, rising affluence, and technological hegemony—all made possible by cheap fossil fuels. The gravity, exigencies, and societal implications of global warming simply are incompatible with the dominant worldview. For conservatives, looking at global warming and other planetary threats through the lens of the dominant worldview is like viewing an apocalyptical science fiction movie. Existential ecological threats are fictional exaggerations; thus, aggressive environmental protection measures are pointless. However, no worldview can protect conservatives—or anyone—from the realities of global warming and our countless other serious environmental threats.

Like seeing storm clouds on the horizon, conservatives glimpsed the threat posed by the environmental movement and the emerging alternative ecological worldview decades ago. In response, they have thrown their energies into building a fortress of denial, concocted partisanship, and political intransigence to protect the fossil-fuel driven economic prosperity essential to their vision of the good life. With the realities of global warming and climate change undermining the foundation of their fortress, conservatives look to solutions compatible with the dominant worldview such as geoengineering and tree planting to address global warming. Unfortunately, the dominant worldview offers a warped lens that distorts the reality of environmental threats and blinds them to the dangers ahead, including the dangers posed by geoengineering.

After decades of denial, duplicity and intransigence, conservatives' support for geoengineering instead of emission reductions is perverse and dangerous. During the thirty-plus years of their falsehoods and political obstruction, global warming has been accelerating and the impacts of climate change have become more severe. After the hottest summer on record at the time, in 2023 *Bioscience* published a report warning that global warming has pushed the earth into "uncharted territory."[78] We need to reduce greenhouse gas emissions faster than earlier IPCC projections. In April 2022, the IPCC stressed, "The evidence is clear: the time for action is now. . . . The decisions we make now can secure a livable future."[79] We do not have the luxury of delay. With global warming accelerating and climate change growing more perilous, pressure is mounting to geoengineer our way to safety.

Do we have time to wait for research and development of successful large-scale carbon capture technologies to come to fruition? Will we have blown past the carbon budget and overshot 2 degrees Celsius of warming long before carbon capture—arguably the safest form of geoengineering—becomes technologically and commercially viable on the scale needed to drawdown atmospheric levels of carbon? The relative simplicity and low cost of injecting aerosols into the stratosphere to cool the planet makes it a tempting alternative to carbon capture and emission reductions. If we deploy untested, large-scale solar geoengineering in desperation, we could wreak havoc on the global climate and ourselves. What ecological, human health, and geopolitical costs and surprises will we incur by continuing the extraction and combustion of fossil fuels? What consequences are we willing to endure and bequeath to future generations to perpetuate the myths of the dominant worldview? Will we condemn our children to living "under a white sky"?

EIGHT

BRIDGING THE PARTISAN DIVIDE

A central tenet of the dominant social paradigm is that the "so-called" ecological crisis has been greatly exaggerated. This worldview has anesthetized adherents with the mythological claim that our superior technological expertise enables us to control and manage a planet created for human prosperity. In sharp contrast to this fantasy of life on earth, in 2021, a multi-national group of environmental scientists published an article titled, "Underestimating the Challenges of Avoiding a Ghastly Future." The scientists issued a sober warning: "future environmental conditions will be far more dangerous than currently believed."[1] Our societal ability to avoid this "ghastly future" ultimately depends on uprooting the dominant social paradigm.

Environmental philosophers, environmental and sustainability studies scholars, and environmental sociologists have long argued that the root of our environmental crisis is our dominant societal worldview. Decades ago, legendary conservationist Aldo Leopold articulated the foundations of a new ecological worldview grounded in the interconnected relationships between humans and all other species in his famous essay, "The Land Ethic."[2] In 1977, shortly after the birth of the environmental movement, E. F. Schumacher noted that "faith in modern man's omnipotence is wearing thin."[3] Similarly, David Ehrenfeld criticized many of the assumptions built into what we now call the dominant social paradigm. Ehrenfeld challenged the hubris and limits of technology, especially in the context of irreplaceable natural resources and irreversible ecological losses. He wrote, "there is little or no chance that the humanists will be able to "engineer" a future . . . resources are approaching their limits—some like topsoil are essential to life and are absolutely non-replaceable."[4] More recently, works like Robin Wall Kimmerer's *Braiding Sweetgrass: Indigenous Wisdom, Scientific Knowledge and the Teachings of Plants* and the scholarship on indigenous environmentalism have painted vivid portraits of lives enacted according to a worldview based on kinship with and respect for all of creation.[5] These works and many others have critiqued assumptions embedded in the dominant worldview including human exemptionalism, unrestrained resource exploitation, the fallacy of infinite economic growth, and the dangers of blind faith in technology.

Unfortunately, it is extremely difficult for people to see the fallacies of the dominant worldview. The view of the world embedded in the dominant worldview simply translates to "reality," the ways things are, the ways things should be. The dominant worldview offers an appealing story of human exemptionalism and invulnerability on a stable, resilient planet. Yet, in spite of the invisibility, sweeping influence, and comfort of the dominant worldview, many people have in fact realized that their perceptions of the environment do not align with the dominant paradigm. The emergence of the new ecological paradigm and subsequent research demonstrate that large numbers of Americans have gravitated toward an alternative environmental worldview.

Uprooting the dangerous dominant worldview demands the twin forces of public pressure and scientific evidence that contradicts the dominant worldview. This potent pairing could drive a potential paradigm shift in American society. In other words, we need the force of a bipartisan environmental movement armed with the battering ram of environmental science to break down the walls of conservatives' fortress of denial and partisan division. We already have half of this equation. After decades of rigorous research by thousands of scientists worldwide, a truly staggering body of scientific evidence clearly—and terrifyingly—demonstrates that the dominant worldview has brought humans to the brink of looming planetary catastrophe. How can we overcome more than three decades of science denial and political distortion to restore the authority of environmental science in environmental politics? How can we shatter the stereotypes of environmentalists and jumpstart the engine of environmentalism? This chapter examines barriers and opportunities for conveying scientific realities and reaching across the partisan gulf to rebuild and rejuvenate a bipartisan environmental movement.

BARRIERS

As architects of denial and division, one of conservatives' most insidious accomplishments has been politicizing the environmental movement and its scientific foundations. A growing number of Republicans have turned away from environmentalism and environmental issues. Gallup notes that prior to 1999, there were no discernible differences based on political party in response to the question, "Do you consider yourself an environmentalist?" Today half as many Republicans self-identify as environmentalists when compared to Democrats.[6] Perpetuating a public discourse about environmental issues that foregrounds controversy and polarization, paired with caricatures of environmentalists, serves the interests of the conservative denial machine that has warped environmental politics in the United States.

Research tells us that paradigms can shift in the face of overwhelming contradictory information. However, partisanship reinforces the distortions and lies about environmentalists and environmental science trumpeted by influential conservative elites, many of whom are widely respected public figures. Many people believe the information manufactured by conservative elites based on party loyalty and because it is consistent with the dominant worldview, a worldview that maintains that a world tailor-made for humans can successfully be managed through unparalleled human technological mastery. Many people also have been persuaded by the dominant worldview's foundational beliefs accentuated by conservatives that environmental "extremists" and "alarmist" scientists have exaggerated or fabricated our environmental problems; thus, no amount of environmental science is persuasive. For Christian evangelicals who believe in their bones that God controls the earth, it is the height of human arrogance to suggest that humans could change the global climate, part of the Almighty's creation. For the political and religious faithful, environmental science is nonsense. These public narratives of partisan division on a variety of environmental issues have led the American public to believe that people are hopelessly divided. For many Americans concerned about climate change, biodiversity loss, toxics, and plastic pollution, this story of polarization has cemented the idea that minds are made up and has caused widespread despair.

Manufactured Partisanship and Lived Experience

Abundant evidence demonstrates that accurate environmental science voiced by scientists and environmentalists has not been able to overwhelm the dominant worldview. The power of scientific information has been negated by the thirty-five-year campaign of duplicity by conservatives intent on delegitimizing the environmental movement and environmental science. But what about people's actual, lived experiences? The vivid scars of habitat destruction and the increasingly severe impacts of climate change are making it tough to escape the physical realities of the world. Ordinary people—even climate change skeptics and ardent denialists—cannot avoid the realities and consequences of capricious weather, debilitating heatwaves, home-devouring wildfires, biblical deluges and massive floods, hurricanes on steroids, shifting growing seasons and new insect pests, and skyrocketing home insurance premiums for those fortunate enough to still have home insurance. In 2021, Yale researchers reported a "dramatic increase" in public worries about climate change after a "brutal year of extreme weather events."[7] They also found that about 50 percent of Americans report that they have personally experienced the impacts of climate change. The deep-rooted belief that the earth can absorb all manner of environmental abuse while remaining a hospitable home for humans is being mightily tested by the inescapable realities of climate change.

No one, including American conservatives, can dodge the reality of climate disruption. However, worldviews shape people's interpretation of events and experiences. For many, the interpretation of their own lived experiences of climate disruption—filtered through the lens of the dominant worldview—has been reinforced by the barrage of climate change denial messages from conservative elites and bolstered by manufactured partisanship. Conservative defenders of the dominant worldview have convinced many that climate change is a naturally occurring phenomenon. Moreover, the dominant worldview leads people to the conclusion that the global climate system is "favorably inclined towards humans . . . even with humans altering the global atmosphere."[8] Researchers and pollsters repeatedly have found that conservatives are less likely to believe that humans are causing global warming and are less worried about climate change impacts. Many conservatives believe that our climate is changing due to natural causes, however chaotically; they argue there always have been extreme weather events. Some Christian evangelicals view environmental destruction and global warming as phenomena that will speed up the coming apocalypse, a welcomed event for the faithful.

Cognitive Predispositions

People's interpretations of their lived experiences are shaped by the dominant worldview and reinforced by cognitive biases. Climate change communication researchers have examined the relationship between worldviews and cognitive predispositions that shape acceptance or rejection of new information including climate change science. Analysts have highlighted the role of "motivated reasoning" in global warming acceptance or denial. Andrew Hoffman explains:

> We interpret and validate conclusions from the scientific community by filtering their statements through our own worldviews. . . . We search for information and reach conclusions about highly complex and politically contested issues in a way that will lead us to find supportive evidence of our pre-existing beliefs.[9]

In other words, as Rebecca Solnit observed, "Given a choice between their worldview and the facts, it's always interesting how many people toss the facts."[10]

People sift and sort, winnowing out information including scientific facts that are inconsistent with their worldview and corresponding beliefs. Like a rocket on a faulty trajectory that bounces off the earth's atmosphere, the massive weight of scientific evidence about the reality of human-caused global warming and other forms of environmental devastation—even when people have experienced climate change and environmental loss—simply bounces off people's heads. Social media algorithms reinforce motivated reasoning, fortifying a world of self-selected information. The power of the dominant worldview and the cognitive biases that reinforce it are amplified by conservatives' parallel effort to delegitimize environmentalists. Clearly, anything those crazy, tree-hugging radicals have to say is nonsense.

OPPORTUNITIES

Does the foregoing suggest we are doomed to remain in the grips of a societal worldview hurtling us toward planetary catastrophe? Although the dominant worldview is maintained by society elites, we cannot overlook the part that public opinion, interest groups, and effective political mobilization play in the dynamic process of social change. An alternative environmental worldview emerged because large numbers of Americans recognized the reality of environmental threats. How can we help people understand the validity of environmental science? How can we support people in a journey away from the dysfunctional DSP to a worldview grounded in ecological realities? What are the opportunities and catalysts for change?

In spite of the hegemony of the dominant worldview in shaping society, politics, and even individuals' interpretations of personal experiences, communication researchers are finding ways to chip away at the foundations of the dominant worldview. Researchers have identified information contradictory to the dominant worldview and conservatives' partisan messaging that overrides cognitive biases, environmental science denial, and partisanship. Carefully selected information can help people overturn mistaken conceptions about environmental issues such as global warming, an important step in the journey toward a new ecological worldview. Understanding the scientific consensus on anthropogenic global warming as well as techniques to strengthen the social consensus provide tools for uprooting the dominant worldview. The combination of discordant information that challenges the dominant worldview paired with social support can be potent catalysts for change.

The Scientific Consensus

Although conservatives have been deviously effective in countering the clout of environmental science, the fact remains that science is powerful and important. We cannot give up on science. Many people frustrated about climate change inaction are certain that sharing the scientific facts is the key to public persuasion: "If people only knew" is a common lament. However, abundant evidence—including the experiences of many novice climate change communicators—shows that sharing the facts can backfire. As we have seen, people resist information that they disagree with. Inconvenient facts bounce right off via motivated reasoning. Nevertheless, research shows people are willing to accept new facts that conflict with deeply held opinions if presented with certain kinds of information. In other words, the *kind* of information matters. A barrage of science or nightmare scenarios about the impacts of climate change is not persuasive. The information that matters is understanding the scientific consensus.

Evidence suggests that when people understand the scientific consensus on global warming, they often adjust their views on global warming and climate change.

Unfortunately, many people are confused or ignorant about the scientific consensus on global warming, causing what researchers call a consensus gap. In 2021, researchers with the Yale Program on Climate Change Communication found that only 25 percent of Americans understand that more than 90 percent of climate scientists agree that global warming is human caused;[11] now 100 percent of climate scientists agree that global warming is caused by humans. Why is this crucial fact not widely recognized? Do people not know the meaning of the word "consensus"? Do people not grasp the differences between a "consensus" and a "majority"?

The scientific consensus on global warming often is framed as a majority opinion in the media. People regularly hear statements about the "overwhelming majority of climate scientists" or the "vast majority of climate scientists." However, there are crucial numerical and political differences between a majority and a consensus. Consensus means there is unanimous agreement, and, by definition, majority means that there is a minority. When people hear the phrase, "the majority of climate scientists agree that humans are causing global warming"—or even the "overwhelming" or "vast majority" of climate scientists agree—what size minority do they envision? What numbers do people see in their heads? What assumptions do they make? As a hypothetical example, if people translate "overwhelming majority" to "80 percent of climate scientists agree," they logically would conclude that 20 percent of climate scientists do not agree that humans are causing global warming. Given the societal implications of addressing climate change, a critically thinking person probably would want to hear what the 20 percent of scientists that disagree have to say. Clearly, distinguishing between the meaning of "majority" and "consensus" is crucial in communicating about climate change. Again, 100 percent of climate scientists agree that human-caused global warming is a real and serious problem; there is no minority opinion.

However, the primary reason that most Americans do not understand the scientific consensus on anthropogenic global warming is due to the decades-long, systematic campaign to distort climate change science in order to deliberately confuse the American public about the realities of global warming. Recall that Frank Luntz, the consultant who advised the incoming George W. Bush administration on environmental policy, emphasized, "Voters believe that there is no consensus about global warming within the scientific community. Should the public come to believe that the scientific issues are settled, their views about global warming will change accordingly."[12] Shaun Elsasser and Riley Dunlap's analysis of conservative columns published from 2007 to 2010 showed that the assertion "there is no scientific consensus" was the most commonly repeated claim about global warming.[13] Conservative denialists have long understood the power of the scientific consensus.

Numerous scholars have pursued research to explore whether understanding the scientific consensus is a significant factor in accepting contested science, including climate change science. Sander van der Linden has described comprehending the scientific consensus on global warming as a "gateway belief" that leads to altered views about climate change:

> The GBM [Gateway Belief Model] postulates that a change in the public's perception of the scientific consensus on an issue acts as a "gateway" to changes in other important cognitive and affective judgments that people may hold, such as the degree to which people think climate change is real, human-caused, and how much people worry about the issue.[14]

In other words, understanding the scientific consensus on global warming is a "gateway" that can lead climate skeptics to a new understanding of climate change. People can change their mind about the reality and dangers of global warming. As Matthew Goldberg et al. emphasize, "when people update their estimates of the scientific consensus, it reflects a genuine update in their beliefs."[15]

Researchers also have examined the relationship between perceptions of global warming, the scientific consensus, and worldviews. Several studies have found that free-market conservatives can change their views on global warming when they understand that there is a scientific consensus on human-caused global warming. When exposed to the reality of the scientific consensus, Goldberg et al. reported, "We also find that conservatives updated their consensus estimates significantly more than liberals."[16] Similarly, in a large-scale replication of the gateway belief model, van der Linden et al. found that "conservatives and climate change disbelievers were more likely to update their beliefs toward the consensus."[17] In other words, as Stephan Lewandosky et al. reported, "Consensus information . . . neutralizes the effect of worldview."[18]

The realization that climate scientists unanimously agree that humans are causing dangerous, global warming—in other words, understanding the scientific consensus—overcomes motivated reasoning and offers a tool to prune the roots of the dominant worldview. Even staunch conservatives who oppose the idea of global warming can change their minds. Frank Luntz was right about the American public: "Should the public come to believe that the scientific issues are settled, their views about global warming will change accordingly."[19] Extinction denial is emerging as the latest form of conservatives' environmental science denial. Communicating the reality of the scientific consensus on the staggering rate of species extinctions also will be a critical communication tool to counter scientific misinformation about biodiversity loss.

Building the Social Consensus

For many Americans concerned about the environment, the story of polarization has cemented the idea that minds are made up and has caused widespread despair. Believing that public opinion on climate change is durably divided, people are afraid to talk about it; many are convinced that climate change is too politically fraught for polite conversation. Yale researchers report that Americans "rarely" or "never" discuss global warming with family and friends.[20] The architects of denial have silenced the American people on one of the most consequential issues we have ever faced as a nation and global community.

The partisan divide manufactured by conservative elites has indeed driven a wedge in the American electorate. From the perspective of the growing number of Americans deeply concerned about climate change and environmental degradation, denying the reality of climate change is unfathomable. It is not uncommon to hear ad hominem criticisms of climate deniers and skeptics. The frustration directed at individual citizens is misguided. Climate denial or confusion is not a measure of character. Climate change denial flows from a faulty worldview bolstered by the decades-long campaign to discredit the environmental movement. Many rank-and-file Republicans seem apathetic about the environment because they have been told it doesn't matter by conservative opinion leaders.

More fundamentally, many see no reason to worry about environmental problems and climate change because they unknowingly are subscribing to the myths of the dominant worldview. If you are oblivious to the fact that the dominant worldview shapes your interpretations of social and physical reality, how are you to know you've been duped? If you never question the premise that the world was designed for human welfare, if you believe that nature can absorb and adapt to all impacts of industrial society, and if you assume we can fix any problems with technology, why would you worry about the environment? When you've been told again and again that extremist environmentalists and alarmist scientists have exaggerated our environmental threats, why would you pay attention to their claims? Armed with the mantle of human exemptionalism, it is full speed ahead with life and politics as usual. Thus, the first compassionate step in draining the moat of partisanship and rebuilding the bipartisan social consensus for action involves de-personalizing the many negative characterizations of climate change deniers and skeptics. Again, it's not faulty human character; it's a faulty worldview and bad politics.

At the birth of the environmental movement, Americans could see and smell and sometimes even taste air and water pollution. Executives in large cities routinely packed an extra dress shirt if they went out for lunch; they knew they would return with grime on their white shirt collars. A series of environmental disasters put environmental degradation on the front pages of newspapers and in families' living rooms as they tuned into the nightly news. An astonishing number of Americans concluded that our mistreatment of the environment had to stop; society had to change. The speed at which concern for the environment took root in American society was extraordinary. This explosion of environmental awareness and concern gave rise to the alternative environmental worldview. Although the dominant worldview is powerful, at some point physical, sensory reality trumps everything.

Today millions of Americans have realized they are experiencing the consequences of climate chaos now; it is no longer a distant worry. Polls show the majority of Americans actually are concerned about global warming; and the number of Americans who worry a great deal about global warming keeps going up. And while it is true that polls show Republicans are less concerned about climate change than Democrats, most polls cannot capture complex realities. There is a multilayered, variegated story unfolding at the ground level. Unfortunately, due to a variety of factors including the politicization of the environmental movement and social fears driven by the partisan divide, people rarely talk about global warming. As a result, people often have faulty impressions about others' beliefs about global warming. Yale researchers report that most people significantly underestimate the number of Americans who believe that global warming is occurring. Most people also underestimate support for policies to address climate change. Gregg Sparkman et al. found that an astounding 80-90 percent of Americans in every state and in every demographic assessed in the study "underestimate the prevalence of support for major climate change mitigation policies and climate concern . . . supporters of climate policies outnumber opponents two to one, while Americans falsely perceive nearly the opposite to be true."[21]

This partisan divide manufactured by conservatives has led to what Nathaniel Geiger and Janet Swim refer to as "pluralistic ignorance," an unfortunate phenomenon that leads to "self-silencing."[22] Similarly, according to Yale researchers, the result of these widely shared mistakes about others' climate change concerns leads to a "spiral of silence." They explain, "There is a climate change 'spiral of silence,' in which even people who care about the issue, shy away from discussing it because they so infrequently hear other people talking about it—reinforcing the spiral."[23] The belief that others are less concerned about climate change causes people to stay quiet out of fear of looking silly or less competent in conversations about climate change. Self-silencing serves the interests of the conservative climate change denial machine. Self-silencing fortifies the partisan divide, cripples social and political mobilization, and feeds despair.

Sharing Stories

Imagine a world in which friends and families actually talk about climate change. Researchers report that having close friends and family members who are concerned about global warming enhances belief in climate change. Goldberg et al. point out that perceived "social consensus" among friends and family is a significant factor that influences beliefs about global warming.[24] Notably, perceived social consensus is particularly important to conservatives. Studies have shown that conservatives put a premium on in-group loyalty and conformity. As Goldberg et al. found, "conservatives and Republicans show a stronger desire than liberals and Democrats to adopt and share the same views with like-minded others."[25] In other words, it is important for ordinary Republicans to see that *other* Republicans are concerned about climate change. The conservative elites in the United States who have systemically denied the reality of global warming for decades are opinion leaders among their conservative followers. However, Goldberg et al.'s research suggests that when conservative individuals see that their family and friends are concerned about global warming—when people talk about global warming—the social consensus among family and friends overcomes the climate change denial pushed by conservative elites.

Beyond simply talking about climate change with family and friends, sharing stories about the personal impacts of climate change can shift others' beliefs about the reality and risks of climate change. In one experiment, researchers tested the impact of radio stories about climate change on conservatives' understanding and acceptance of human-caused global warming. They found that radio stories about conservative Republicans who advocate climate change solutions interwoven with traditional Republican themes and values—such as "free markets, pro-life values, national security, and religion"—can impact conservatives' beliefs about global warming.[26] Again, Republicans need to see that *other* Republicans are concerned about climate change.

One story that was particularly effective is the tale of conservative Republican Bob Inglis. Inglis was a multi-term congressman from South Carolina with sterling conservative credentials. His tenure on the House Science Committee led Inglis to publicly declare that climate change is real. Inglis knew this public statement would cost him his seat in Congress; Trey Gowdy defeated Inglis in a primary landslide in 2010. In 2015, Inglis was awarded the JFK Profile in Courage Award for his public stance on climate change. After being trounced in the primary, Inglis embarked on a campaign to demonstrate the feasibility of addressing climate change with policy tools and approaches compatible with conservative values. Inglis is now the executive director of the non-profit group republicEn, home of the EcoRight.[27] As one republicEn member emphasized, "I've never really seen a right-wing party care about the environment before. I'm very passionate about it and I'm very right wing. So thank you."[28]

In other words, when conservatives hear stories about other conservatives concerned about climate change, they often shift their perceptions of global warming. Shared stories about global warming reinforce social norms about the acceptability of concern about climate change among conservative Republicans in defiance of Republican elites' barrage of climate change denial messages. Sharing stories about climate change with family and friends, in organizations like republicEn and in online spaces, strengthens the social consensus essential for political mobilization. Sharing stories chips away at the foundation of partisanship.

Other researchers have stressed the importance of sharing personal stories about the emotional impacts of climate change. Abel Gustafson et al. tested the impact of a fisherman's emotional account of "the loss of beloved wildlife and his own way of life" on others' views about global warming.[29] Gustafson et al. found that personal stories from "relatable individuals" such as conservative hunters and anglers about the emotional impacts of climate change and loss can "shift the beliefs and risk perceptions of political conservatives and moderates."[30] Hunters and anglers often have conservative political views, but climate change and biodiversity loss are having a profound impact on their way of life.[31] Climate change is reducing habitat and harming prized wildlife and fish populations. Changing seasons affect reproductive success, and some species are migrating to new territories. In fact, long before climate change led to observable and measurable impacts on biodiversity, habitat loss and fragmentation driven by agriculture, urban sprawl, and energy development have caused declining populations of many prized game species. According to the National Wildlife Federation and Conservation Science Partners, game species have lost "on average, 6.5 million acres of vital habitat over the last two decades."[32]

A little-known story about Rachel Carson's *Silent Spring* involves reactions to the book by outdoorsmen. Although a review in *Time* magazine accused her of having a "mystical attachment to the balance of nature," outdoorsmen were sympathetic to Carson's work because it matched their own personal experiences.[33] According to Livia Gershon, "Men in rural areas described 'dead' rivers and the destruction of wildlife by aerial DDT sprayings."[34] After reading *Silent Spring,* many of these outdoorsmen connected their concerns with a more comprehensive ecological viewpoint. Outdoorsmen's concerns about the environmental effects of widespread spraying of pesticides were reflected in Ed Dodd's comic strip featuring forest ranger "Mark Trail" in 1963. "Mark Trail" warned that "when man indiscriminately broadcasts chemical insecticides and nuclear wastes in his environment . . . he seriously threatens the future of ALL living things including himself!"[35]

Outdoorsmen, hunters, and anglers have observed the steady erosion of wildlife habitat and species populations for decades. Presently, in the American west, populations of large game species such as antelope and elk are shrinking due to habitat loss. In the northeast and southeast, habitat loss due to suburban sprawl and agriculture has led to the virtual disappearance of formerly common game species such as the Northern

Bobwhite; rural dwellers rarely hear their sweet "Bobwhite" spring song anymore. One observer lamenting their disappearance wrote,

> It used to be a rural way of life! Get up before dawn, eat a plate of ham gravy and biscuits, load the bird dogs in a box on the back of a pickup truck and head to the corn, soybean fields, or bicolor lespedeza and sedge fields. . . . Unfortunately, it's an adrenalin rush and a way of life that children of the new millennium will likely never know.[36]

One lifelong sportsman who has personally experienced the negative impacts of global warming on wildlife populations, Todd Tanner, founded the group Conservation Hawks. Hoping to educate and motivate fellow sportsmen, Conservation Hawks has produced educational films and electronic and print media about the relationship between climate change and biodiversity to protect their sporting heritage.[37] Facebook hosts a Conservative Hunting and Angling Caucus where members post information and concerns about environmental and political threats to their cherished hobbies. During the first Trump administration, conservative sportsmen lobbied against Trump's decision to open up several western national monuments for oil and gas drilling; they sent letters and created ads critical of the proposed redesignation.[38] According to a national poll conducted by the Theodore Roosevelt Conservation Partnership and Public Opinion Strategies, "77 percent of Republicans and 80 percent of Democrats supported preserving the existing national monuments that allow hunting and fishing."[39] For many outdoor enthusiasts, conservation is as important as gun rights.[40]

Politically conservative hunters and anglers can be powerful communicators about the realities and impacts of habitat destruction and biodiversity loss, especially important as extinction denial grows in prominence. Biodiversity loss has not been politicized to the same degree as climate change; with the right approach, it can be safer to discuss. Habitat loss and ecological degradation are eroding many people's quality of life and cherished hobbies, causing grief and solastalgia. Sharing personal stories about the emotional reality of habitat and biodiversity loss compounded by climate change strengthens the social consensus essential for political mobilization. In other words, talking about how biodiversity loss affects ordinary people like politically conservative hunters and anglers and other outdoor enthusiasts can be a backdoor into conversations about other environmental issues including climate change.

Entering Conversations Through the Back Door

What does it mean to enter a conversation about climate change—or any controversial issue—through the back door? Conversations about global warming and the extinction crisis often backfire due to individuals' entrenched positions on the issues. When positions are challenged, people typically dig in to defend their position. It can be difficult for people to change their positions without losing face. In our culture, changing your mind often is viewed as a weakness. We deride politicians who change their positions as flip-floppers, forgetting that changing one's mind often is a sign of education and personal growth.

However, there is a crucial difference between people's stated positions and their underlying interests. Years ago, dispute resolution theorists and practitioners, Roger Fisher and William Ury, wrote a landmark book, *Getting to Yes: Negotiating Agreement Without Giving In.*[41] In this slim volume, they laid out the principles of interest-based negotiation, a style of negotiation that differentiates between people's positions and interests. Positions are people's publicly stated beliefs or demands; interests are the underlying issues that people care about. People tend to argue over positions; however, people typically have a variety of interests underlying their stated positions. This negotiation method has been utilized in countless situations ranging from interpersonal conflicts to workplace disputes to large, multi-party environmental disputes. In virtually all of these conflict situations, participants believed that resolution was impossible. Resolution often is impossible when people focus on their positions. When positions are challenged, conflict often escalates as people bang heads.

Yet, in countless conflict situations, dispute resolution practitioners have demonstrated that it is possible to circumvent entrenched positions and forge agreements based on people's underlying interests. For example, years ago Tallahassee, Florida, was attempting to site a new urban parkway. Forty organizations including local business interests, state and local government, universities, environmental groups, and neighborhoods were involved in the controversy. Environmental groups took a firm public position in opposition to the new parkway. Thinking citizen involvement would reduce the conflict, planners asked environmental stakeholders if they would participate in the siting decision for the parkway. If the environmentalists had agreed to participate in the siting decision, they would have had to walk away from their publicly stated position; building the parkway was baked into the problem definition. The environmentalists said no, and the conflict escalated as the partisans continued to fight over deeply held, clashing positions. Eventually, an environmental mediator was brought in.[42] She asked the environmental stakeholders if they would participate in a consensus process to discuss Tallahassee's transportation needs. The environmentalists said yes. Talking about transportation needs opened the door to discussing the environmental stakeholders' interests: they wanted to discuss ways of reducing automobile travel such as creating carpool and bike lanes and improving public transit in order to obviate the construction of the new parkway. There are countless examples like this in the files of environmental mediators and dispute resolution practitioners.

When it comes to conversations about climate change, public opinion polls and the research on "self-silencing" tell us that far more people are amenable to talking about it than most people realize. Nevertheless, conservatives' successful politicization of global warming has caused widespread perceptions of entrenched positions on climate change. As a result, people are afraid to talk about it, afraid to provoke an argument. The politicization of climate change has turned the issue into a Godzilla-like monster towering over our social and political lives, sowing fear and stomping out conversation. However, it is untenable to give up on conversations about global warming. In our families, in our communities, and in society, we simply cannot remain silent about global warming, an issue that literally poses an existential threat to modern civilization.

Given the fear of tortuous conversations about global warming, what can we do? It is essential that we remember the difference between interests and positions. Just because someone says they don't think global warming is human-caused or it's not a serious problem, that doesn't mean that the same individual doesn't care about a host of issues and concerns affected by climate change and an economy powered by fossil fuels. *People have multiple interests.* We don't have to tackle stated positions on global warming head-on. By focusing on interests, we can enter conversations about climate change through the back door. The partisan gulf carefully crafted by conservatives has caused people to lose sight of American's longstanding love of our natural heritage. By sharing stories about nature, we can sidestep politicized positions that divide us. Americans all across the political spectrum, including political conservatives, traditionally have identified with the values of conservation. As one young member of republicEn lamented, "I hate that the fact that I am conservative automatically labels me as not caring for the planet. Since when does respecting all life in every form make you a member of either political party?"[43]

One story in particular—the history of the Arctic National Wildlife Refuge—offers a powerful example of Americans' bipartisan, enduring love of our natural heritage. The federal system of national parks, forests, and wildlife refuges was created during the administration of Teddy Roosevelt, a prominent Republican. In 1960, President Eisenhower established the 8.9-million-acre National Arctic Wildlife Range in Alaska for the purpose of "preserving unique wildlife, wilderness and recreational values."[44] In 1980, President Carter signed the Alaska National Interest Lands Conservation Act, which renamed the "National Arctic Wildlife Range" the "Arctic National Wildlife Refuge" (ANWR) and expanded the size of the refuge to over 19 million acres.[45]

President Reagan tried to open ANWR for oil and gas drilling, but the attempt failed. Shortly after Reagan's proposal ran aground, the 1989 *Exxon Valdez* oil spill in Alaska's Prince William Sound had a powerful effect on the politics of ANWR. Public reactions to nightmare images of oil-soaked wildlife put the proposal on hold for years. The George W. Bush administration resurrected proposals to drill in ANWR. If there ever were a time that Americans would have supported increased domestic energy production, the months after the terrorist attacks of 9/11 would have been ripe with opportunity. The Bush administration had successfully convinced Americans that Iraq—an oil-producing nation—had perpetrated the attacks. But like his predecessors who tried to open ANWR, the Bush administration's plan to open ANWR failed.

Although ANWR is a spectacularly beautiful place rich in wildlife, it is so remote that most Americans are unlikely to ever see it. Yet, in response to the Bush administration's efforts to open ANWR, in March of 2002— five months after the 9/11 terrorist attacks—a Gallup poll revealed that only 36 percent of Americans supported drilling in ANWR. A bare majority of Republicans—51 percent—favored oil drilling in ANWR; 64 percent of Independents and 68 percent of Democrats were opposed.[46] When the first Trump administration opened ANWR via a provision tucked into the Tax Cuts and Jobs Act of 2017, researchers reported that two-thirds of Americans still were opposed to drilling in ANWR.[47] Yale researchers found that "The number of voters who are strongly opposed to drilling in the ANWR outnumber those who strongly support the policy by nearly four to one."[48] About 81 percent of Democrats, 64 percent of Independents, and 50 percent of Republicans opposed drilling in the ANWR; a mere 17 percent of Republicans strongly supported the policy. The story of ANWR illustrates that Americans do not even need to personally experience our national treasures in order to support their protection. In response to public pressure, on his first day in office President Biden suspended the ANWR oil and gas drilling leases authorized during the first Trump administration.[49]

Conserving our natural heritage offers fertile ground for initiating conversations about climate change and biodiversity loss through the back door of interests. As our history illustrates, protecting our public lands—state and national parks and forests, wildlife refuges, and national monuments—is a widely shared interest among the American people. As reservoirs of wildlife habitat, our public lands are home to beloved iconic species such as the bison and moose of Yellowstone National Park, the alligators and flamingoes of Everglades National Park, and the migrating herds of caribou in the Arctic National Wildlife Refuge. Our public lands provide recreational opportunities and breathtaking scenery enjoyed by millions. According to the camping app Dyrt, 80 million people in the United States are campers; more than half have annual incomes under $100,000.[50] Research conducted by the non-profit group Save Our Winters found that 50 million Americans—including hikers, climbers, surfers, bikers, skiers, and trail runners—look to the outdoors for most of their recreation time. Perhaps more importantly, Save Our Winters found that the majority of outdoor enthusiasts—regardless of political affiliation—believe the federal government needs to do more to address climate change.[51] Protecting our public lands is important for a variety of reasons that can be discussed with or without referencing global warming or climate change or environmentalists.

Although pollsters tell us that we have a yawning partisan gap over environmental issues, there is abundant evidence to the contrary. For example, a majority of Americans of all political persuasions supported the Biden administration's 30/30 plan to protect 30 percent of America's land, ocean areas, and inland waters by the year 2030.[52] According to one poll, 72 percent of Republicans, 90 percent of Democrats, and 68 percent of Independents supported the plan.[53] In an age when policies often are rejected by members of the public based on nothing more than the party affiliation of the program sponsor, the bipartisan support for the Biden's administration's 30/30 plan is striking: 72 percent of Republicans supported the plan. Similarly, a 2023 poll by the National Parks Conservation Association (NPCA) found evidence of strong bipartisan support for wildlife conservation. The NPCA surveyed public opinion on a host of issues connected to wildlife conservation including climate change, land development, wildlife corridors and habitat connectivity, ocean plastics, overfishing, and air pollution.[54] They found that "85% of Republicans and 91% of Democrats believe more needs to be done to protect national park wildlife." Perhaps more significantly, they found that "74% of Republicans, 89% of Democrats and 85% of Independents believe that climate change threatens national park wildlife."[55] It bears repeating: 74 percent of Republicans worry about the effects of climate change on wildlife.

The NPCA poll also found that 82 percent of Americans support reducing or eliminating the sale of single-use plastics such as plastic water bottles and straws in our national parks in order to protect marine wildlife.[56] The conservative *Washington Times* has published a variety of articles on plastic pollution; a 2022 article highlighted a study by Beyond Plastics that found greenhouse gas emissions from plastics will overtake coal-fired plants in the United States by 2030.[57] Similarly, a 2018 article in the conservative *Washington Examiner* titled "Conservatives Should Support Action to Reduce Plastic Pollution" argued that "government must take the lead to protect the public good."[58]

Entering through the back door of interests marked "wildlife" or "plastic pollution," sharing personal stories and concerns about wildlife conservation and single-use plastics builds on Americans' traditional love of nature and offers pathways to broader conversations about biodiversity loss and climate change. Sharing stories about wildlife conservation, outdoor recreation, and the societal plague of plastics also offers opportunities for weaving some environmental science into conversations in a non-confrontational way. More importantly, talking about environmental issues with others across the partisan divide fortifies the bipartisan social consensus foundational to energizing an effective social and political movement for change.

In fact, there are multiple back doors into conversations about climate change. Talking about widely shared American values and concerns—focusing on interests—can lead to productive conversations. For example, Americans across the political spectrum care about national pride, patriotism, and economic competitiveness. Do we want to cede U.S. economic and technological leadership to China or any other world economy? America still could be the global leader of the twenty-first-century economy powered by renewable energy. A 2024 poll by the centrist think tank Third Way found that 77 percent of Americans support investments in manufacturing clean energy.[59] Similarly, a 2019 poll by the Pew Research Center found broad bipartisan support for renewable energy: 62 percent of those who lean Republican and 82 percent of moderate/liberal Republicans support renewable energy. Even among conservative Republicans, almost half (49 percent) support renewable energy.[60]

Can we make room for people to support renewable energy without directly linking the topic back to global warming? If people's interests align with the goals of decarbonization, does it matter if they don't think global warming is a major threat? Renouncing a stated position opposing the reality and seriousness of global warming can be difficult for people. However, with patience, entering through the back door of interests labeled "economic competitiveness" can lead to conversations that eventually flow from the benefits of renewable energy for U.S. global economic leadership—as well as for consumers—to the topic of global warming. With or without stated concerns about global warming, support for renewables moves us as a society in the necessary direction away from fossil fuels. The time is ripe to seize on this opportunity. Not that long ago, there were few accessible alternatives to fossil fuels. With the price of renewables tumbling, access to clean, affordable renewable energy is within reach for many Americans. Tax credits and other incentives built into the Inflation Reduction Act will further expand access to renewable energy options for many Americans.[61]

The story of climate change policy adoption in the states is illustrative of the popularity of renewable power even in the face of climate change denial and polarization. Barry Rabe's research on state climate policies reveals that many states have adopted mechanisms such as renewable portfolio standards, green power offsets, and financial incentives to encourage the adoption of renewable power in the name of economic self-interest.[62] For example, although Texas Governor Greg Abbott aggressively protects Texas's oil and gas industries, Texas is now the nation's largest producer of solar and wind power.[63] In 2022, Iowa's GOP Platform referenced "alleged global warming," and Governor Kim Reynolds has said that the effects of climate change are "overstated";[64] nevertheless, wind energy now powers roughly 57 percent of Iowa's energy needs.[65] In 2023, Jason Grumet, CEO of American Clean Power, noted,

> Iowa is a clean energy success story, producing and using clean power for a greater share of the state's electricity generation mix than any other state. Thanks to the leadership of Senator Grassley and Governor Reynolds, Iowa has achieved this success with broad public support.[66]

And even Florida—a state whose governor has ordered the removal of the term "climate change" from state statutes—led the nation in solar installation in 2023.[67]

Rabe explains that states like California that face severe climate change impacts have adopted state policies explicitly framed as a response to environmental threats. However, policy entrepreneurs in some states have framed state climate policies as opportunities for economic development, sometimes carefully avoiding use of the terms "climate change" or "global warming."[68] For example, Indiana now ranks fourth in the nation in terms of renewable energy capacity development. Former governor of Indiana Eric Holcomb believes there are "multiple causes" of global warming, and he praised President Trump's decision to withdraw from the Paris Climate Agreement. Nevertheless, in 2022, he led an economic development trip to the UN climate conference held in Egypt (COP 27) to deliver a speech about Indiana's commitment to renewable energy. As *Politico* noted, "his presence at the conference marks a new moment in the battle against climate change, as economic development opportunities in green energy lure unlikely converts."[69]

Conservatives increasingly are recognizing the economic benefits derived from renewable power. The Conservative Energy Network (CEN), a "national network of state-based organizations focused on promoting clean energy innovation rooted in conservative values," focuses on market-based approaches to securing the green energy economy.[70] In 2022, CEN convinced Florida Governor Ron DeSantis—proponent of "don't say climate change" laws—to veto a bill that would have restricted residential rooftop solar installations in Florida.[71] In other words, for many states, renewable energy development aligns with economic self-interests regardless of whether state officials acknowledge the reality and dangers of global warming. By looking at climate policy in the states, we see that focusing on interests is not only a productive conversation tool; policy entrepreneurs that strategically focus on state interests have been able to advance meaningful climate change policies under the guise of economic development. Again, instead of tripping over positions on global warming, we need to remember our collective interests in decarbonization.

We all pay a huge price for using ancient forms of carbon as a source of energy. Our reliance on coal, oil, and gas carries countless social and environmental costs that Republicans, Democrats, and everyone in between care deeply about. We can explore the collateral damages of our reliance on fossil fuels without even mentioning global warming or climate change. Fossil fuel extraction, refining, shipping, and combustion have consequences for geopolitics and national security, protection of wild places and biodiversity, private property rights, urban air quality, public health, environmental justice, and the proliferation of plastic pollution. We also can capitalize on existing public advocacy directed at the collateral damages of fossil fuels without directly confronting positions on global warming. For example, in countless communities with soaring asthma rates, parent groups have been lobbying against diesel-powered school buses to protect their children from harmful air pollution. Weary of the sight of plastic bags fluttering in trees like grotesque national birds combined with growing worries about the health consequences of microplastics, people in communities all across America are lobbying against single-use plastics. Remarkable alliances of conservative farmers and landowners, conservative private property rights advocates, veterans, millennials, and Native Americans are challenging the expansion of pipelines for fossil fuels and captured carbon dioxide. If large numbers of Americans' lives are being compromised by our reliance on fossil fuels, if people advocate—and vote— for ending the fossil fuel era, does it matter if some individuals are not persuaded that global warming is a human-caused, serious problem? Talking about the costs of a world powered by fossil fuels strengthens the social consensus on the need for change without charging in through the front door and challenging positions on climate change.

The sustainability movement provides another back door into conversations about climate change and other environmental issues. Despite some right-wing critiques, sustainability is a popular and powerful concept in contemporary American life. Sustainability integrates environmental, social, and economic concerns; thus, it provides multiple pathways into conversations about the environment even for people who do not naturally gravitate to environmental issues. For example, the social and economic dimensions of sustainability offer doorways into conversations about social justice, economic equity, sustainable economic growth, and economic competitiveness. For people interested in the corporate sector, many large firms have been at the forefront of the sustainability movement. For example, 3M has been a sustainability leader for decades. Beginning with their "Pollution Prevention Pays" initiative in 1975, 3M demonstrates the evolution of corporate sustainability goals originally focused on reducing environmental impacts to the broader adoption of "Environmental, Social and Governance" (ESG) goals encompassing a wide array of waste reduction, greenhouse gas emission reduction, and diversity and social justice goals and strategies.[72] Today, some of the world's largest corporations have adopted rigorous sustainability ESG goals including Verizon, Apple, Microsoft, Cisco Systems, and Bank of America.[73]

The economic dimensions of sustainability also can lead to conversations about the growing importance of the green power sector. The success of the Inflation Reduction Act (IRA) provides a powerful illustration of the essential role of investments in green technology to spur job growth and achieve our national climate goals. According to the World Resources Institute, the IRA has

> unleashed a manufacturing renaissance . . . nearly doubling the amount of manufacturing construction in just one year, with forecasts of even higher growth in years to come. Since its passage, makers of battery components, wind and solar equipment, and electric vehicles have announced tens of billions of dollars of new investments, bringing significant local opportunities with them.[74]

For the many Americans weary of partisan struggle and gridlock at the national level, sustainability can lead to conversation and engagement with important issues at the community and household level. Local sustainability programs have drawn attention to the importance of reuse, repair, waste reduction, recycling, composting, food choices and food waste, single-use plastics, and mass consumption. In many communities, thrift stores help combat the environmental impacts of fast fashion and reduce clothing expenditures. Libraries of Things enable people to borrow household items such as sports equipment, sewing machines, tools, baby gear, and children's toys without having to purchase them. Libraries of Things ease pressures on home finances and reduce personal consumption, an often-overlooked source of greenhouse gas emissions given that we outsource many of our purchasing-related emissions to manufacturers in China. Community sustainability plans and initiatives also have expanded access to home weatherization, renewable energy, and alternative forms of transportation including mass transit, safe bike paths, and electric vehicle charging stations—all of which reduce fossil fuel consumption.

Food waste is a major sustainability issue that offers many opportunities for bipartisan conversation. The USDA reports that Americans waste 30-40 percent of the food supply.[75] Discarded food squanders precious water resources as well as large amounts of petroleum-based fertilizers, pesticides, and the fossil fuels to power farm equipment and transport food to suppliers and markets. Project Drawdown emphasizes that eliminating food waste at all points in the supply and consumption chain can lead to meaningful emission reductions.[76] According to the Pew Research Center, 80 percent of Americans say they consciously work to reduce food waste to protect the environment.[77] The Columbia Climate School newsletter shared an illustrative nonpartisan story about food waste: when a young woman told a friend who was attempting to cram leftovers into her freezer stuffed with food scraps destined for the farmers market composting station, her friend exclaimed, "Wow . . . that's not very *Republican* of you."[78] Entering through the back door of reducing food waste invokes widely shared interests including grocery budgets, food security, natural resource use, and climate change.

Thus, there are multiple pathways for overcoming the forces of denial, transcending partisanship, and undermining the foundation of the dominant worldview. We need to remember that in a world of fake news and alternative facts, carefully chosen, accurate information still matters. Understanding the scientific consensus can change minds. As extinction denial gains momentum, exposing people to the robust scientific consensus on the scale and gravity of biodiversity loss also will be an important communication strategy.

Talking about the scientific consensus on both issues puts people on a path toward acceptance of the environmental science that contradicts the dominant worldview and the falsehoods told by its conservative defenders. And contrary to the story of intractable political division, countless Americans are concerned about a variety of environmental issues including climate change, biodiversity loss, plastic pollution, air pollution, public health, and sustainability. Americans all across the political spectrum treasure our natural heritage. Talking about public lands and wildlife conservation builds on this enduring foundation and opens up space for broader conversations about biodiversity loss and global warming. Focusing on the human and environmental collateral damages of a world powered by fossil fuels opens another wide doorway into conversations about climate change. Conversations about community sustainability can transcend partisanship, build social cohesion, and lead to conversations about new models for sustainable living that enhance quality of life and offer alternatives to traditional fossil fuel-dependent lifestyles.

Climate scientists assert that the most important thing people can do to combat global warming is "talk about it."[79] We must have the courage to heed their advice. It is impossible to mobilize a social movement around issues that people are afraid to discuss. We might ask, in the post-*Dobbs* era, what would be the fate of the reproductive choice movement if people were afraid to talk about abortion? It is essential that we break the "spiral of silence" and start talking to each other about environmental issues. Discussing climate change, biodiversity loss, plastic pollution, sustainability, and other environmental issues important to people strengthens the social consensus on the need for political change. We all need to talk about global warming and biodiversity loss with family and friends, sharing our fears, experiences, anxieties, and grief over disappearing green space and lost outdoor recreational opportunities. By talking with each other about our collective environmental interests and concerns, strengthening the social consensus on the reality and shared consequences of environmental problems, we can drain the partisan moat that defends the fortress of denial created by conservative defenders of the dominant worldview. By talking with each other, we can create the momentum for political change. If we simply have the gumption to talk to each other, one by one, story after story, together, we can write a new story of environmental politics. Focusing on widely shared interests, we can tiptoe in through the back door and ignite a bipartisan movement for social change.

ENVIRONMENTALISM AND DEMOCRACY

What if regenerating a bipartisanship environmental movement had broader implications for American society? Could the social cohesion of a bipartisan movement help us rewrite the fractured story of contemporary American politics? Examining the social and economic effects of climate change suggests potential answers to these questions. The International

Committee of the Red Cross warns that climate change "may indirectly increase the risk of conflict by exacerbating existing social, economic and environmental factors."[80] The U.S. military, the United Nations, and security experts agree that global warming is a "threat multiplier" that will intensify societal disruptions, economic losses, and global unrest.[81] However, climate change *already* is multiplying the economic losses and social disruption caused by extreme weather and wildfires here at home. Climate change has caused billions of dollars in losses due to damaged infrastructure, crop and livestock destruction, real estate losses, rising insurance claims, and overstretched public services. By November 2024, the United States had sustained twenty-four confirmed weather/climate disaster events with losses exceeding $1 billion in 2024 alone.[82]

Repeated global warming–fueled hurricanes, fires, and floods are bankrupting many small communities.[83] Homeowners in states prone to climate change–related disasters are increasingly finding home insurance out of reach. In states like California, Florida, and Louisiana, it is nearly impossible for some homeowners to obtain property insurance; many others can no longer afford soaring premiums.[84] Some homeowners living in high-risk areas already have lost their private home insurance and have turned to state-backed insurance programs.[85] According to the *Tampa Bay Times,* "a flood of new policyholders are joining state-backed insurance 'plans of last resort,' leaving states to assume more of the risk on behalf of residents who can't find coverage in the private sector."[86]

Climate change-driven migration and societal disruption, once a topic confined to the global south, is a growing reality in the United States. Headlines in popular news sources proclaim, "Climate Migration Has Come to the United States" and "Climate Change Will Force a New American Migration."[87] Wildfires, hurricanes, and droughts pose existential threats to communities, particularly in the west and southeast. In 2023, wildfires burned 2.6 million acres, destroying thousands of homes; and fire seasons are starting earlier and growing longer. Hurricane season also is growing longer and more dangerous; Hurricane Beryl made history by becoming the first Category 5 storm on record in July: on July 1, 2024. Hurricanes are becoming more destructive as the energy of global warming fuels catastrophic winds, torrential rains, and record storm surges that cause massive flooding and property destruction, transforming whole communities into debris piles of pick-up sticks.

Some of the most popular cities in the United States—New York City,

New Orleans, Boston, and Miami—will be transformed as rising sea levels displace millions. Even now, it floods on sunny days in some southern Florida coastal communities. According to a *Forbes* magazine survey, "Nearly a third of Americans surveyed cited climate change as a reason to move in 2022."[88] CBS News reports that about 3 million Americans already are climate migrants.[89] Abrahm Lustgarten, author of *On the Move: The Overheating Earth and the Uprooting of America*, warns that existing trends will only grow worse. He writes,

> The United States will be rendered unrecognizable by four unstoppable forces: wildfires in the West; frequent flooding in coastal regions; extreme heat and humidity in the South; and droughts that will make farming all but impossible across much of the nation.[90]

Already, extreme heat increases the risk of suicide among American farmers.[91]

In 2021, *The Lancet* published a warning from the editors of more than 200 medical journals worldwide. Noting that the "science is unequivocal," they cautioned that global warming and biodiversity loss pose the risk of "catastrophic harm to health that will be impossible to reverse."[92] In the summer of 2023, wildfires raging across Canada sent smoke over the border, triggering air quality alerts in eighteen states from Montana to New York and as far south as Georgia; the alerts lasted for days, forcing many Americans indoors. Wildfire smoke causes respiratory illnesses, and people who are frequently exposed to wildfire smoke may have higher risks of developing dementia later in life.[93] Global warming-fueled "heat domes" brought scorching temperatures to much of the country for days on end in 2023. The searing heat of 2023 caused discomfort and dangerous conditions for millions, especially for outdoor laborers, the urban poor, and others living in substandard housing. Nighttime temperatures also remained stubbornly high; according to the 2018 National Climate Assessment, nights are warming faster than days.[94] High nighttime temperatures magnify the dangers of heatwaves, placing a "particularly heavy burden on the body, raising the risk of heat illness and death."[95]

Millions of Americans are living with often unacknowledged and poorly understood emotional and economic anxieties, physical discomfort, and health problems driven by global warming. What are the effects of climate destabilization on American society? Is global warming a "threat multiplier" here at home? Research has shown that violent crime and online hate speech soar during heatwaves.[96] According to a study in *The Lancet*, "Online hate speech on Twitter was 22% higher in extreme heat with temperatures between 107 and 113 degrees Fahrenheit." The researchers found that online hate speech increased significantly at both the city and state level regardless of climate zone, income levels, and religious and political beliefs.[97] During the summer of 2023, Phoenix had fifty-four days of temperatures above 110 degrees including a thirty-one-day streak of temperatures at or above 110 degrees.[98] The NOAA weather forecast app turned red for cities in Texas, Louisiana, and Mississippi as temperatures soared above 107 degrees.[99] Some midwestern cities sweated through heat indices well above 107 degrees; in late August, the heat index was 134 degrees in Lawrence, Kansas.[100] According to the Anti-Defamation League, "Online hate and harassment surged in the 2023 findings . . . a third of American adults (33%) experienced some form of online harassment in the past twelve months, up from 23% in 2022."[101]

What happens when the seasons no longer match our memories of seasons past? During the summer of 2023, scorching temperatures made visits to some of our national parks almost impossible. The mercury hit 109 degrees twice in Moab, Utah, gateway to the popular Arches National Park;[102] Death Valley shattered global records with a nighttime temperature of 120 degrees.[103] Toxic algal blooms closed beaches all over the country and as far north as Wisconsin, Vermont, and the Pacific Northwest.[104] Ocean waters off the coast of Florida felt like a hot tub, 101 degrees. Coral reefs bleached, and many species of fish—including freshwater fish in overheated inland lakes, rivers, and streams—sought relief in deeper, cooler waters.[105]

In our warming world, *Climate Central* warns, "anglers will struggle to plan fishing trips that connect with fish that are biting."[106] When you can't spend summer vacation exploring our national parks without risking heat exhaustion, and you can't count on a successful summer fishing trip or a relaxing day at the beach, playing in the cooling waves, what do you do? When the earth beneath your feet literally is changing in disconcerting and often dangerous ways, who do you blame? Who do you look to for leadership, sympathy, or solace? Could finding allies across the political spectrum within the environmental movement build the momentum for bipartisanship more broadly throughout society? Could a shared commitment to protecting our natural heritage revitalize Americans' commitment to defending our democratic heritage? Could it be that "Conserving the country's natural resources—land, air and water—is patriotic"?[107]

NINE

TRUTH, LIES, AND HOPE

We cannot successfully tackle the daunting challenges of our journey through the Anthropocene if we are blinded by falsehoods and robbed of the nourishment of hope. Authentic hope is grounded in honesty. Building the social consensus on the need for action on climate change, biodiversity loss, and other planetary threats requires both *scientific* and *political* honesty. Authentic hope grounded in honesty about our planetary perils is an essential first step for people dealing with anxiety and grief caused by living in our world of wounds. Hope grounded in honesty about environmental politics is crucial to combatting frustration and despair and mobilizing action. Offering an honest account of environmental politics harnesses the power of emotion in politics. Exposing the offenders and their lies unleashes the productive energy of anger that cuts through the emotional inertia that cripples political mobilization. This concluding chapter delves into the power and potential of truth, lies, and authentic hope for rebuilding a bipartisan environmental movement.

PARADIGMS AND PARTISANSHIP REVISITED

Environmental politics is fundamentally a story about clashing paradigms, not simply competing partisans. Throughout this book, I have often referred to "conservative defenders of the DSP" or simply "conservative elites" rather than to political parties. The original researchers who identi-

fied the dominant social paradigm and the new ecological paradigm argued
that analyzing environmental politics based on the traditional left-right
ideological spectrum is too one-dimensional. The conventional left-right
continuum excludes the earth. Both liberals and conservatives endorse
policies consistent with the DSP worldview. For example, most Republican
and Democratic politicians typically prioritize economic growth over environmental protection. Both parties have supported policies that expand
fossil fuel extraction, production, and infrastructure, including miles of
pipelines. And both parties have endorsed geoengineering research, including solar radiation management, the most controversial form of geoengineering.

Regardless of whether we look through the lens of conventional partisan politics or the lens of competing paradigms, the Democratic Party
must share some of the blame for our current environmental dilemmas,
including inaction on climate. In his 2021 book *They Knew: The US Government's Fifty-Year Role in Causing the Climate Crisis*, Gus Speth provides
a chronology of bipartisan federal inaction on climate change.[1] Speth calls
out Democratic administrations from President Carter up through the
Obama administration for their support of fossil fuel expansion and their
anemic responses to global warming.

Reluctant Democrats may have taken cover under Republicans' climate change denial and avoided taking tough political positions on global
warming. Persuaded by conservatives' campaign of disinformation, some
Democrats may have believed the science of global warming was too uncertain to justify a federal policy response. Democratic politicians are just
as indebted to campaign donors and their own vested interests as Republicans. And Democrats are just as likely to use various strategic maneuvers
to weaken, delay, or prevent the adoption of policies that undermine their
own political interests. As a prime example, witness coal baron Joe Man-
chin, former Democratic senator from West Virginia, using every trick
up his sleeve to stymie the Biden administration's climate agenda—every
political maneuver except lying about the reality of anthropogenic global
warming. And that is the difference.

Conventional accounts of environmental and climate change politics,
firmly grounded in the norm of political neutrality, provide painstaking
analyses of both parties' actions and failures. However, typical accounts
of environmental politics succumb to the fallacy of false equivalency. The
Democratic Party's failures in the realm of environmental and climate

change politics do not match the scale of the Republican Party's purposefully crafted strategy of distortion and lies to block meaningful action on a host of environmental issues, including climate change. Although individual Democrats—like all politicians—are not above stretching the truth, the Democratic Party has not challenged the science of biodiversity loss, acid rain, and ozone depletion, and they have not lied to the American people about the reality and dangers of climate change. They have not partnered with the fossil fuel industry and contrarian scientists to purposefully manufacture scientific uncertainty and lie about the scientific consensus on anthropogenic global warming. The Democratic Party has not tried to delay action on global warming by claiming that alarmist scientists and environmental extremists are exaggerating the threats of climate change. They have not harassed the climate scientists sounding the alarm. The Democratic Party has not supported geoengineering research after decades of climate change denial and deceit as a substitute for aggressive emission reductions in order to continue our reliance on fossil fuels. And the Democratic Party has never sought to undermine the environmental movement. In fact, Republican politicians and conservative pundits have inextricably linked the Democratic Party with environmentalism to politicize and discredit them both. An article published in the conservative paper the *Washington Times* in 2023 accuses Democrats of "obsessive bowing to their god of environmentalism."[2]

Traditional political analyses apportion equal blame to both parties. How were Democratic administrations supposed to enact and successfully implement aggressive and enduring environmental and climate change policies without good faith Republican partners? How was the Democratic Party supposed to overcome the fortress of political and partisan opposition so carefully cultivated by the rearguard Republican champions of the DSP worldview, defenders of unrestrained exploitation of earth's resources and guardians of political and economic fortunes soaked in oil? In recent years, a growing number of individual Republican legislators have acknowledged the reality of climate change and the need for a political response. This trend is particularly evident among young conservatives, many of whom have joined the EcoRight group republicEn. However, the Republican Party remains staunchly opposed to mitigating global warming by leaving behind the fossil fuel era. It is time to openly acknowledge the reality and consequences of Republican elites' political maneuvers in the realm of environmental politics. As fierce, unscrupulous defenders of the DSP, conservative Republican elites have contested and contorted the environmental movement and environmental science and have systematically lied to the American people about global warming for over three decades.

However, the purpose of this book is not simply to call out Republican elites. The purpose of this book is to tell the story of how worldviews shape perceptions of reality and the extreme measures powerful interests will take to protect the dominant worldview. It is the story of how conservative elites have lied about environmental and climate change science to defend their political and economic fortunes inextricably grounded in the DSP worldview. It is a story of how conservatives have worked to delegitimize the environmental movement—the public face of the alternative ecological worldview—in the eyes of the public. And it is a story of how conservatives have manufactured a partisan divide over environmental issues that separates Americans from each other and their love of nature. It is time to tell this story.

AUTHENTIC HOPE IN A WORLD OF WOUNDS

Like most young people, our students are grappling with growing up in a world of wounds. Their confidence in their ability to achieve the traditional benchmarks of adult success—productive careers, home ownership, a family, economic security—has been shaken by a world rocked by environmental, political, and economic turmoil. Environmental damage and climate change are major sources of worry and anxiety as they stare into a future of ecological impoverishment, climate instability, and disruption. Research shows that concerns about the state of the planet are interfering with young peoples' "sleep, their ability to study, to play, and to have fun."[3] How do we sustain younger generations condemned to coming of age in a world of wounds? By speaking truth to those who will be in power.

Truth and Hope

As an educator teaching environmental and climate change politics, I have sustained myself and my students by cultivating authentic hope. Authentic hope is built on staring reality in the face, and then consciously choosing hope and action. We cannot fortify ourselves with authentic hope if we live in a world of "alternative facts" and lies. Authentic hope is "rooted in unalloyed reality."[4] Students and members of the public don't need blind optimism, reassuring words, or false promises. We need authentic hope rooted in the truth about the world. Building authentic hope begins with an unflinching look at the reality of our planetary woes. Many of my students are overwhelmed by the severity of biodiversity loss, climate change, and our other environmental problems. Many are shocked by the extent of the damage already occurring and terrified by visions of future environmental loss and climate chaos. One semester, a senior shared her dismay that she was about to graduate from college and had never learned about the state of the earth in any previous college course. Most are relieved to put a name to the vague sense of grief—solastalgia—they have felt living in a world of wounds but not understood. They are relieved to know they are not alone as they grapple with feelings of loss and sadness.

Nurturing authentic hope is a process that requires emotional honesty and courage. It is a journey that involves acknowledging grief and despair on the road toward authentic hope. Suzanne Moser and Carol Berzonsky explain:

> [F]ostering authentic hope in ourselves . . . begins with having the courage . . . to go to hopelessness, to our own despair . . . grieving actual and anticipated losses. . . . [W]e have to be continually willing to go back into grief and come back out.[5]

Coping with environmental loss and grief is similar to grieving the death of a loved one. Steven Running, University of Montana climate scientist and Nobel Prize recipient for his work on the 2007 IPCC Fourth Assessment Report, developed "The 5 Stages of Climate Grief" modeled after Elisabeth Kübler-Ross's pioneering work on the process of grieving.[6] Just like mourning the loss of a beloved family member or friend, climate grief involves "Denial, Anger, Bargaining, Depression and Acceptance."[7] Authentic hope is built on the last stage: acceptance of the realities of a world of wounds paired with action.

As a first step in their journey toward authentic hope, my students write about the ecological losses they have experienced. They talk to family and friends about the emotional impacts of climate change, biodiversity loss, and other environmental threats. I share my own stories of environmental grief and loss, and my own passage from despair to authentic hope. I share my own experiences of solastalgia, like the grief I felt when yet another global warming-fueled windstorm brought down the huge ash tree and the uncommon orchid that had sheltered at its base. I tell them about

the fracking leasing frenzy in my rural community that threatened my
sense of home and safety. I share how I had nightmares about my teenage
daughter learning to drive, forced off narrow, winding country roads by
huge fracking trucks; and nightmares of my dog poisoned by toxic fracking fluids. After weeks of worry, I remembered Jerome Groopman's wise
words: "Because nothing is absolutely determined, there is not only reason
to fear but also reason to hope. And so we must find ways to bridle fear
and give greater rein to hope."[8] I accepted that fracking could profoundly
alter my rural home, but I recognized that fracking had not yet come to
my backyard. Faced with fear and hope, I chose hope. By choosing hope,
I regained my balance and my ability to teach more effectively. I encourage my students to recognize that hope is a conscious choice we can make
every day. Together, we stare reality in the face, and then we choose hope.
Pushing back against a media landscape dominated by chaos and controversy, we actively seek out positive stories about actions taken to address
climate change and other environmental problems. Throughout the semester, we revisit the meaning and importance of authentic hope as the
basis for action.

Building authentic hope also requires learning the truth about environmental politics. The conventional bipartisan story of both political parties'
failures to adequately address climate change and other environmental
problems is grounded in false equivalencies and dysfunctional political
norms. This story simply causes more frustration and despair. Telling the
real story of environmental politics means stepping out from behind the
curtain of academic and political norms of nonpartisanship. Telling the
real story means telling the truth about conservative Republican elites'
carefully crafted edifice of environmental science denial and political obstruction. Telling the real story of environmental politics means taking a
deep dive into their distortions and lies.

My students investigate the many tentacles of the climate change
denial movement. Believing the partisan gulf in environmental politics is
real and immutable, my students typically enter my course assuming that
most Americans are unconcerned about global warming. "Why don't they
care?!" is a familiar refrain. Ironically, learning that the American people
have been lied to for over thirty years about the reality and dangers of
global warming is reassuring. They realize the problem isn't that people
are just greedy or selfish or stupid or unconcerned. I tell my students, "If
members of your family don't accept the reality of global warming—if they

told you global warming isn't real—remember, they have been lied to. We all have been lied to." Staring political realities right in the face—learning that Americans have been lied to for decades—helps my students build authentic hope.

Anger and Hope

Fostering authentic hope involves confronting all the emotions evoked by global warming and climate change politics, including grief, despair, and anger. My students are dismayed and angry at the audacity of the lies spun so smoothly by the Republican guardians of the DSP worldview, defenders of the fossil fuel industry. I urge my students to acknowledge and use the energy of their anger. A 2021 study showed that anger over climate change inaction is an effective strategy for overcoming climate change anxiety and depression; it also propels people into activism. As Samantha Stanley et al. emphasize, "experiencing eco-anger predicted better mental health outcomes, as well as greater engagement in pro-climate activism."[9] Anger defeats despair and can motivate people to be involved in the climate change movement.

Perhaps more importantly and surprisingly, another study demonstrated that conveying public anger about climate change inaction can bridge the partisan divide. Anandita Sabherwal et al. found that public messages that express people's anger over U.S. climate change inaction helped people realize that others are worried about climate change and clamoring for action. Again, we see the power of the social consensus; people need to see that others are concerned—and frustrated—about global warming inaction. Perhaps even more significantly, Sabherwal et al. found that messages about shared anger over climate change inaction heightened demand for social change across partisan groups.[10] Amidst all the ideas about how to overcome the partisan divide, who would have guessed that messages about shared anger over climate change inaction might bring us together?

We all should be angry. Republican elites and the fossil fuel industry—conservative defenders of the dominant worldview—have lied to the American people, to Democrats, Independents, and rank-and-file Republicans alike. We all have been lied to about the reality, dangers, and solutions to climate change. We've been exhorted to *Reduce your carbon footprint!* using a "carbon calculator" strategically created by BP to shift the responsibility for global warming to consumers.[11] We've been fed ludicrous lies about recycling and renouncing plastic drinking straws as solutions to global warming. We've been blamed for our "addiction" to fossil fuels even though accessible, affordable energy is essential for going about the business of life. Are we also "addicted" to housing or employment? In 2024, Darren Woods, CEO of ExxonMobil, claimed, "The world isn't on track to meet its climate goals—and it's the public's fault."[12] Ordinary people aren't the drivers of global warming; even the most carbon-neutral lifestyles barely make a dent in total emissions. ExxonMobil, Shell, and about a hundred other fossil fuel companies are responsible for almost three-fourths of all greenhouse gas emissions.[13] The textbook authors who claimed "The villains of the climate change story are ordinary Americans" were dead wrong.[14] Climate scientist Michael Mann once said that deliberate climate change denial constituted the "most villainous act in the history of human civilization."[15] Conservative Republicans—and their allies in think tanks, media, and the fossil fuel industry—are the real villains in the climate change story.

CALLING OUT THE VILLAINS

In 1988, climate scientist James Hansen warned Congress that global warming had arrived. Although leaving behind the fossil fuel era poses complex political, economic, and technological challenges, if the United States and the global community had acted thirty-five years ago, many of the worst impacts of climate change we already are experiencing could have been averted. What share of the blame for our global predicament should we assign to the Republican leaders and conservative pundits in the United States who have systematically and repeatedly lied about the dangers we face? Analysts have observed, "The Republican Party stands alone in its conviction that no national or international response to climate change is needed."[16] What share of the blame should we lay on the world's only major climate-denialist party? What kind of world would we be living in now if both political parties in the United States had embraced the reality and science of global warming?

According to NASA, global CO_2 emissions reached record levels in 2023—the hottest year on record.[17] Global average temperatures reached 1.5 degrees Celsius warming—the ceiling agreed upon by the world's nations—several times in 2023-2024. What if the United States—the world's largest, historic emitter of greenhouse gases—had adopted the mantle of leadership in climate diplomacy thirty years ago? Would global CO_2 emissions have fallen to safer levels? Would the transformation of the electric grid be well underway? Would we have been able to prevent climate chaos?

Admittedly, international negotiations over an issue as complex as global warming were never going to be easy and uncomplicated even with unwavering U.S. leadership. Failures and frustrations were inevitable. David Victor points out "when the global warming problem first appeared on their radar screens the world's top diplomats opened a toolbox that had all the wrong tools for the job."[18] For example, there is widespread agreement that the 1997 Kyoto Protocol, the first international global warming treaty, was fatally flawed. Victor argues that Kyoto was doomed largely because it relied on a global emissions trading market worth at least $2 trillion, an unprecedented creation of assets by voluntary international treaty.[19] The Kyoto Protocol also divided the world into developed and developing nations, an artificial division in the context of emerging economies like China and India with growing greenhouse gas emissions.[20]

Experts spent almost two decades analyzing the failures of Kyoto and the subsequent years of unproductive climate negotiations. After years of failed climate diplomacy, Victor argued for a "radical rethinking" of international climate negotiations. Victor asserted that a variety of more flexible bottom-up approaches at the national, regional, and global levels were more likely to succeed than top-down agreements. Victor recommended a "climate club" consisting of a small group of nations that matter most to our climate future, paired with useful commitments based on realistic government efforts rather than emission outcomes.[21] Victor's assessment aligns with observations made by Rafe Pomerance, the deputy assistant secretary of state for environment and development during the Clinton administration. After his experience with the Kyoto Protocol, Pomerance emphasized, "Climate change policy is not fully dictated by negotiations. . . . They're dictated by domestic political opportunities and constraints."[22]

The architects of the Paris Agreement listened to the many post-Kyoto critiques and recommendations. Given opportunities provided by the falling price of renewables, they framed emission reductions as a "race to the top" for leadership in the clean energy economy of the twenty-first century. Grounded in national self-interest, emission reductions were framed as an opportunity rather than a burden. Renouncing globally binding treaties and mandatory emission reductions, the Paris Agreement was based on voluntary "nationally determined contributions" in the context of a "pledge and review" framework.[23] In other words, each nation determined its own feasible level of emission reductions and agreed to participate in periodic reviews of progress. What if both Democrats and Republicans had embraced the Paris Agreement, seizing the opportunity for the United States to lead the twenty-first-century energy economy powered by U.S. technological innovation and investment?

Most analysts concur that the Paris Agreement was a major achievement and important first step. However, the global community is far from meeting the goals of the Paris Agreement. On November 15, 2024, a group of scientists and climate diplomacy experts sent an open letter to all members of the Conference of the Parties (COP) calling for an update on how the COP functions.[24] Among others, their recommendations included preventing nations who are not seriously committed to transitioning away from fossil fuels from participating in the COP, streamlining the process of negotiations to put greater emphasis on delivering specific actions, and strengthening the role of science to better incorporate the latest scientific findings.[25]

It is possible that a comprehensive, top-down approach to global climate negotiations never will succeed. Charles Sabel and David Victor have characterized international agreements like the Paris Agreement as, at best, an umbrella under which local partnerships between business and governments can facilitate experimentation in order to push the technological frontier necessary for combatting global warming.[26] However, there is an important, overlooked question we must ask: was it realistic to expect a robust global response to global warming when the world's largest historic emitter of greenhouse gases has been such an unreliable partner for decades? Admittedly, other industrialized nations also put a high priority on fossil fuel-powered economic growth. But what if the United States had acknowledged the gravity of unchecked global warming and cast itself as the global leader in climate diplomacy similar to our role in international security? What if the United States—the "city on a hill" illuminated by renewable energy—had led the global transition to carbon-free energy? With unwavering United States leadership, would the global community still be facing climate chaos and devastating economic losses?

The United States has seesawed between global climate leadership and withdrawal. President George H. W. Bush campaigned on the seriousness of global warming, pledging to use the "Whitehouse effect" to combat the "greenhouse effect"; as president, he signed the United Nations Framework Convention on Climate Change in 1992. However, by the end of the George H. W. Bush administration, the Republican Party had thoroughly embraced the strategies of doubt and denial.[27] The Clinton administration helped broker the landmark Kyoto Protocol. After rejecting the Kyoto Protocol, the George W. Bush administration did not try to resuscitate global negotiations, and domestically, the administration denied and suppressed climate science. The Obama administration played a key role in the success of the Paris Agreement by pursuing bilateral negotiations with China prior to the Paris climate talks. The first Trump administration withdrew the United States from the Paris Agreement. On his very first day in office, President Biden signed the instrument for the United States to rejoin the Paris Agreement, and the Biden administration made significant commitments to reducing domestic greenhouse gases.

However, most Republicans were vociferously opposed to the Biden administration's climate policies. Even though many red states are the recipients of generous funds allocated under the Inflation Reduction Act (IRA), not a single Republican voted for the legislation and some red states have refused to accept funding from the IRA. And after pledging to withdraw the United States from the Paris Agreement during his 2024 campaign, on January 21, 2025, President Trump withdrew the United States from the Paris Agreement once again.[28] Trust is essential to successful negotiations. How could any other nation trust the United States' enduring commitment to negotiations and emission reductions? Without the leadership and good faith commitment to progress from the United States, why would other nations aggressively reduce their greenhouse gas emissions?

We cannot ignore the domestic and international implications of Republican leaders' decades-long campaign of lies, distortions, and obstruction. As it stands now, after more than thirty-five years of lies, the Republican Party claims their constituents aren't clamoring for action on climate change. What a clever boomerang of deception and betrayal. What would Americans voters have demanded if they had been told the truth?

A CHORUS OF TRUTH

Admittedly, referring to environmental and climate change science as "truth" goes against scientific, religious, and cultural norms. Scientists would be the first to assert that their scientific findings do not constitute truth. However, in a complex modern society, perhaps truth should refer to more than religious and philosophical formulations of the concept. Decades of environmental and climate change research by thousands of scientists—countless peer-reviewed studies conducted by tireless investigators all over the world seeking knowledge and professional success rather than personal monetary gain—have produced a powerful and compelling scientific consensus on global warming, biodiversity loss, and other major planetary threats. Faced with irreversible environmental impacts and climate chaos that literally will disable modern civilization if left unchecked, surely it is time to refer to the scientific consensus on global warming, biodiversity loss, and other global environmental hazards as the truth about our planetary predicament. As David Joravsky observed, "The world's scientific communities cannot claim absolute truth, but they can fairly claim that they are closer than anyone else to genuine knowledge concerning their particular fields of study."[29]

Telling the truth about environmental and climate change science is not without obstacles. The politics of science in the United States are complicated and contested. People are confused about a host of scientific issues other than climate change such as vaccine safety, nutrition, and GMO food. The Republican Party's self-serving lies about climate change have become deeply enmeshed in broader cultural narratives about science reinforced by social constructionism. Republican politicians and pundits transformed the scientific consensus on climate change, biodiversity loss, and other planetary threats into socially constructed, political accords that buttressed their politicized distortions of environmental science and the environmental movement.

Nevertheless, we have to push back against the forces that undermine environmental science. We have to push back against the lies and "alternative facts" about our planetary reality. Decades ago, Rachel Carson ignited a revolution. Grounded in science, her landmark book *Silent Spring* sent shockwaves through the American public and mobilized the large middle class. Captivated and horrified by Carson's eloquent, painstakingly accurate portrayal of the scientific case against the indiscriminate use of DDT, Americans rose up and demanded protection of the environment. Factual, scientific information undimmed by orchestrated forces of denial carried the new science-based, environmental movement forward into the hearts and minds of the American public and the halls of Congress. Scientific knowledge was transformative.

However, telling the truth about environmental and climate change science is only half of the equation. We also must tell the honest story of environmental politics. We need to tell the story of the deliberate, dangerous duplicity of the Republican Party and its allies even though partisanship and motivated reasoning create hurdles to belief. No doubt, ardent rank-and-file Republicans will never accept that their leaders have lied to them about a host of environmental issues including global warming. Although the forces of climate change denial are powerful and ever evolving, we must tell the honest story of environmental politics. We can tell the honest story fortified by the realization that the conservative denial machine has purposely constructed an artificial edifice of partisanship that warps the reality of Americans' shared environmental concerns. It truly is astounding—and promising—that 80-90 percent of Americans underestimate the degree of public concern and support for climate change policies among the public. As Gregg Sparkman et al. observed, Americans truly are experiencing "a false social reality."[30] All across the political spectrum, many more people than we realize are concerned and are willing to listen.

We must tell the truth about environmental politics because this is the missing story. There are many accessible sources of accurate climate change science. For example, meteorologists are one of the most trusted voices of science; many of them are talking about global warming and its effects on weather, even at the risk of professional retaliation. After sharing stories about the relationship between extreme weather events and climate change, Chris Gloninger, the chief meteorologist for CBS affiliate KCCI-TV in Des Moines, Iowa, resigned after receiving harassing emails and death threats that caused PTSD and family health problems.[31] Climate change science is complicated and hard for most people to comprehend. Grasping the reality of climate change politics is much easier. People don't need advanced science degrees to understand the story of the deliberate denial and deceit of the fossil fuel industry and conservative elites. People don't need a science background to comprehend how conservatives have denigrated and caricaturized environmentalists; the many stereotypes about radical tree huggers are all too familiar.

We must tell the truth about environmental politics in order to activate the power of emotion. Emotions play a major role in politics, shaping political communication, behavior, and mobilization. Exposing the lies about global warming can enable us to capitalize on the productive energy of anger. Anger can fuel a bipartisan environmental movement strong enough to overwhelm the forces of denial and despair. As Stanley et al. observed, "Our findings implicate anger as a key adaptive emotional driver of engagement with the climate crisis."[32] Finally, we need a chorus of truth about climate change politics to nourish authentic hope. We owe it to the worried children whose lives are being robbed of innocence and carefree days and the anxious young adults staring into a future of climate chaos and lost opportunities. And we owe it to the millions of people across the globe whose lives have been devasted by unchecked global warming, robbed of the moral and political leadership of the United States.

CONCLUSION

There was a time when Americans were motivated by science and love of nature to rally in defense of the environment. The nightmare evoked by Rachel Carson's "silent spring" devoid of bird songs horrified and motivated millions of Americans who flocked to the fledgling environmental movement. What would our world look like today if Americans had been given honest scientific information about the gravity of ecological destruction and global warming years ago? What would our world look like if conservative elites had not disparaged the environmental movement, purposely driving people away from the activists leading the way to an environmentally sustainable future? Would accurate environmental science and a vital, bipartisan environmental movement paired with public outrage and love of the environment have propelled a societal shift to the NEP worldview?

The fundamental challenge in environmental politics—and for society at large—is confronting the dangerous delusions of the dominant social paradigm. The dysfunctional worldview fiercely defended by conservative elites is driving the world into rampant environmental ruin, irreversible ecological losses, and planetary instability. In the shadow of this clash of worldviews, our children's lives and the fate of future generations loom large. Challenging this worldview demands telling the honest story of environmental science and politics. It means summoning the courage to jettison conventional academic and political norms and offering a frank account of conservative elites' long-running campaign of deceit. It means telling the story of how American conservatives have methodically manipulated environmental science and public opinion about the environmental movement, lying to the American public about the realities of environmental problems—including the existential threats of climate change—for decades.

Some might worry that abandoning political norms of neutrality would aggravate the partisan gulf. Ironically, telling the truth about environmental politics is the pathway to bipartisanship and authentic hope. Conservative elites have lied to ALL Americans about the environmental movement and the reality and dangers of our planetary threats. By telling the truth about environmental politics, we can show people that conservative defenders of the dominant worldview have manufactured an artificial partisan gulf to protect a fortress of falsehoods with no durable foundation. We must synchronize all of our voices in a chorus of truth to shatter the "spiral of silence" that maintains conservatives' duplicity. Together we can overwhelm the forces of denial and deceit and propel a shift to a societal worldview grounded in ecological sanity.

The foundation for building the necessary social consensus for political action and social change rests on the enduring, bipartisan appreciation of our natural heritage among the American people. Americans of every political persuasion care about clean air and water, diverse public lands and outdoor recreation, and wildlife of all kinds from backyard birds to singing whales and thundering herds of bison. Building on this lasting love of our natural heritage, families and friends need to talk about climate change and our shared environmental losses. Academics, teachers, newscasters, media personalities, social media influencers, clergy, and civic and organizational leaders of all kinds must seize every opportunity to talk about the environment and climate change. All of us must have the courage to tell the real story of environmental politics. Together, we can unleash the power of truth, anger, and hope. Armed with new tools and perspectives, courage and conversation, together we can jumpstart the engine of environmentalism powered by the shared love of our planetary home.

If we ignore the clash of worldviews playing out before us, we will miss opportunities to change the trajectory of our environmental story. If we simply accept the conventional story of environmental politics—tolerating the false equivalencies and the lies about the environmental movement and science, believing the manufactured partisan gulf is immutable—we will hasten our journey into planetary instability. We will squander our children's and future generations' ecological inheritance.

We still have time to forge a new, bipartisan story of environmental politics. It is not too late to be trustworthy ancestors.

Appendix

The New Ecological Paradigm Scale

Dominant Social Paradigm Statements	New Ecological Paradigm Statements
Humans have the right to modify the natural environment to suit their needs.	We are approaching the limit of the number of people the Earth can support.
Human ingenuity will ensure that we do not make the Earth unlivable.	When humans interfere with nature it often produces disastrous consequences.
The Earth has plenty of natural resources if we just learn how to develop them.	Humans are seriously abusing the environment.
The balance of nature is strong enough to cope with the impacts of modern industrial nations.	The balance of nature is very delicate and easily upset.
The so-called "ecological crisis" facing humankind has been greatly exaggerated.	Despite our special abilities, humans are still subject to the laws of nature.
Humans were meant to rule over the rest of nature.	The Earth is like a spaceship with very limited room and resources.
Humans will eventually learn enough about how nature works to be able to control it.	If things continue on their present course, we will soon experience a major ecological catastrophe.

Source: Riley E. Dunlap, Kent D. Van Liere, Angela G. Mertig, and Robert Emmet Jones, "Measuring Endorsement of the New Ecological Paradigm: A Revised NEP Scale," *Journal of Social Issues* 56, no. 3 (2000): 425-442.

Notes

Preface

1. Nicholas F. Benton, "Einstein: Everything Is Energy," *Falls Church News-Press Online*, October 7, 2021, accessed November 14, 2024, https://www.fcnp.com/2021/10/07/einstein-everything-is-energy/

2. It impossible to tell the rich story of environmental politics from the perspective of one academic discipline. Thus, my work challenges traditional norms of academic specialization and assumptions about expertise. Cultural and professional expectations about expertise typically assume deep knowledge and decades of experience in one field. With advanced degrees in both the natural and social sciences, mine is a multi-disciplinary expertise. My intellectual strength involves integrating across breadth, making connections, and helping a wide audience understand complex social and environmental problems.

3. Johannes Persson et al., "Toward an Alternative Dialogue Between the Social and Natural Sciences," *Ecology and Society* 23, no. 4 (2018), https://doi.org/10.5751/ES-10498-230414

4. Sven Ove Hansson, "Social Constructionism and Climate Science Denial," *European Journal for Philosophy of Science* 10, no. 37 (2020), https://doi.org/10.1007/s13194-020-00305-w

5. "1992 World Scientists' Warning to Humanity," *Union of Concerned Scientists*, July 16, 1992, accessed October 31, 2024, https://www.ucsusa.org/resources/1992-world-scientists-warning-humanity

6. William J. Ripple et al., "World Scientists' Warning to Humanity: A Second Notice," *BioScience* 67, no. 12 (2017): 1026-1028, https://doi.org/10.1093/biosci/bix125.

7. Corey J. A. Bradshaw et al., "Underestimating the Challenges of Avoiding a Ghastly Future," *Frontiers in Conservation Science* 1 (2021), https://doi.org/10.3389/fcosc.2020.615419

8. "The Evidence Is Clear: The Time for Action Is Now. We Can Halve Emissions by 2030," *IPCC Newsroom*, April 4, 2022, accessed October 28, 2024, https://www.ipcc.ch/2022/04/04/ipcc-ar6-wgiii-pressrelease/

9. Raymond Zhong, "These Climate Scientists Are Fed Up and Ready to Go on Strike," *New York Times*, March 1, 2022, https://www.nytimes.com/2022/03/01/climate/ipcc-climate-scientists-strike.html; see also: Bruce C. Glavovic et al., "The Tragedy of Climate Change Science," *Climate and Development* 14, no. 9 (2022): 829–833, https://doi.org/10.1080/17565529.2021.2008855

10. Kai N. Lee, *Compass and Gyroscope* (Island Press, 1993).

11. William Lafferty and Eivind Hovden, "Environmental Policy Integration: Towards an Analytical Framework," *Environmental Politics* 12, no. 3 (2003): 1–22, https://doi.org/10.1080/09644010412331308254

12. Lawrence E. Susskind et al., "Arguing, Bargaining and Getting Agreement," in *The Oxford Handbook of Public Policy*, ed. Michael Moran, Martin Rein, and Robert E. Goodin (Oxford University Press, 2005), 269–295.

13. Walter A. Rosenbaum, *Environmental Politics and Policy* (Sage/CQ Press, 2020).

Introduction

1. Rachel Carson, *Silent Spring* (Houghton Mifflin, 1962).

2. Hazel Erskine, "The Polls: Pollution and Its Costs," *Public Opinion Quarterly* 36, no. 1 (1972): 120–135, https://doi.org/10.1086/267984

3. "Clean Air Act," 42 U.S.C. §7401 et seq. (1970).

4. "Clean Water Act," 33 U.S.C. §1251 et seq. (1972).

5. "Safe Drinking Water Act," Pub. L. No. 93-523, 88 Stat. 1660 (1974).

6. "Endangered Species Act of 1973," Pub L. No. 93-205, 87 Stat. 884, 16 U.S.C. §§ 1531–1544.

7. Frank Newport, "Update: Partisan Gaps Expand Most on Government Power, Climate," *Gallup*, August 7, 2023, accessed October 30, 2024, https://news.gallup.com/poll/509129/update-partisan-gaps-expand-government-power-climate.aspx

8. Alec Tyson and Brian Kennedy, "How Americans View Future Harms from Climate Change in Their Community and Around the U.S.," *Pew Research Center* (blog), October 25, 2023, accessed October 30, 2024, https://www.pewresearch.org/science/2023/10/25/how-americans-view-future-harms-from-climate-change-in-their-community-and-around-the-u-s/

9. David W. Orr, "Optimism and Hope in a Hotter Time," *Conservation Biology* 21, no. 6 (2007): 1392–1395, http://www.jstor.org/stable/4620979

10. Jerome Groopman, *The Anatomy of Hope: How People Prevail in the Face of Illness* (Random House, 2004), xiv.

11. Aldo Leopold, *A Sand County Almanac, with Other Essays on Conservation from Round River* (Oxford University Press, 1966), 197. The full quote by Leopold: "One of the penalties of an ecological education is that one lives alone in a world of wounds" (167).

12. David Malakoff, "'Sink into Your Grief.' How One Scientist Confronts the Emotional Toll of Climate Change," *Science*, April 12, 2021, https://www.science .org/content/article/sink-your-grief-how-one-scientist-confronts-emotional-toll -climate-change

13. Gaia Vince, "How Scientists Are Coping with 'Ecological Grief,'" *The Guardian*, January 12, 2020, https://www.theguardian.com/science/2020/jan/12/how -scientists-are-coping-with-environmental-grief

14. Caroline Hickman et al., "Climate Anxiety in Children and Young People and Their Beliefs About Government Responses to Climate Change: A Global Survey," *The Lancet Planetary Health* 5, no. 12 (2021): e863-e873, https://doi.org/10 .1016/S2542-5196(21)00278-3

15. Richard Schiffman, "Climate Anxiety Is Widespread Among Youth—Can They Overcome It?," *National Geographic*, June 29, 2022, https://www.nationalgeo graphic.com/environment/article/climate-anxiety-is-widespread-among-youth -can-they-overcome-it

16. Quoted in Mélissa Godin, "Eco-Grief Around the World," *Atmos*, December 1, 2021, accessed October 30, 2024, https://atmos.earth/ecological-grief-climate -change-mental-health/

17. "Art Works for Change," accessed October 30, 2024, https://www.artworks forchange.org/portfolio/chris-jordan/; https://www.chrisjordan.com/Midway/1/ thumbs-caption

18. Quoted in Lisa Bennett, "An Abiding Ocean of Love: A Conversation with Chris Jordan," *DailyGood*, accessed October 29, 2024, https://www.dailygood.org/ story/493/an-abiding-ocean-of-love-a-conversation-with-chris-jordan-lisa-ben nett/; see also Anna Turns, "The Photo That Made the Plastics Crisis Personal," *BBC*, June 2, 2023, accessed December 27, 2024, https://www.bbc.com/future/arti cle/20230531-the-photo-that-changed-the-worlds-response-to-the-plastics-crisis

19. Tyson and Kennedy, "How Americans View Future Harms."

20. "Addressing Climate Change Concerns in Practice," *American Psychological Association*, accessed October 29, 2024, https://www.apa.org/monitor/2021/03/ce -climate-change

21. Susan Clayton et al., "Mental Health and Our Changing Climate: Impacts, Inequities, Responses," *American Psychological Association*, 2021, accessed October 30, 2024, https://www.apa.org/news/press/releases/mental-health-climate -change.pdf

22. Kate Schapira, *Lessons from the Climate Anxiety Counseling Booth: How to Live with Care and Purpose in an Endangered World* (Hachette Books, 2024).

23. "Iceland Holds Funeral for First Glacier Lost to Climate Change," *The Guardian*, August 18, 2019, https://www.theguardian.com/world/2019/aug/19/ iceland-holds-funeral-for-first-glacier-lost-to-climate-change

24. Sasha Starovoitov, "Glacier Funerals Offer a Way of Coping with Ecological Grief," *State of the Planet* (blog), Columbia Climate School, September 24, 2021, accessed October 30, 2024, https://news.climate.columbia.edu/2021/09/24/ glacier-funerals-offer-a-way-of-coping-with-ecological-grief/

25. Zach Roberts, "Deniers Rally Conservatives to Dismiss Climate Science as 'Fake News' and Share Breitbart Stories Instead," *DeSmog*, March 2, 2017, accessed October 29, 2024, https://www.desmog.com/2017/03/02/deniers-rally-conserva tives-dismiss-climate-science-fake-news-breitbart-cpac/

26. Clive Hamilton, *Defiant Earth: The Fate of Humans in the Anthropocene* (Polity Press, 2017), viii.

27. Hamilton, *Defiant Earth*, vii.

28. Melinda H. Benson, "New Materialism: An Ontology for the Anthropocene," *Natural Resources Journal* 59, no. 251 (2019), https://ssrn.com/abstract=34 47625

29. Basil Bornemann, "The Anthropocene and Governance: Critical Reflections on Conceptual Relations," in *The Anthropocene Debate and Political Science*, ed. Thomas Hickmann, Lena Partzsch, Philipp Pattberg, and Sabine Weiland (Routledge, 2019), 60.

30. "Andrew Dobson: Trajectories of Green Political Theory," interview by Luc Semal, Mathilde Szuba and Olivier Petit," *Natures Sciences Sociétés* 22, no. 2 (2014): 132-141, https://doi.org/10.1051/nss/2014021

31. Andrew Dobson, *Green Political Thought* (Routledge, 2007); Robyn Eckersley, *Environmentalism and Political Theory: Toward an Ecocentric Approach* (SUNY Press, 1992); John M. Meyer, "Political Theory and the Environment," in *The Oxford Handbook of Environmental Political Theory*, ed. Teena Gabrielson, Cheryl Hall, John M. Meyer, and David Schlosberg (Oxford University Press, 2016), 773-791.

32. Anthony Burke et al., "Planet Politics: A Manifesto from the End of IR," *Millennium: Journal of International Studies* 44, no. 3 (2016): 499-523, https://doi .org/10.1177/0305829816636674

33. Delf Rothe, "Global Security in a Posthuman Age? International Relations and the Anthropocene Challenge," in *Reflections on the Posthuman in International Relations*, ed. Clara Eroukhmanoff and Matt Harker (E-International Relations, 2017), 87-101, accessed October 30, 2024, https://www.academia.edu/34742550/ Global_Security_in_a_Posthuman_Age_International_Relations_and_the_An thropocene_Challenge

34. Nicholas H. Hedlund-de Witt, "Towards a Critical Realist Integral Theory: Ontological and Epistemic Considerations for Integral Philosophy," 2013, accessed October 29, 2024, https://www.academia.edu/4661222/Towards_a_Critical_ Realist_Integral_Theory_Ontological_and_Epistemic_Considerations_for_Inte gral_Philosophy

35. Frank Biermann, "The Anthropocene: A Governance Perspective," *Anthropocene Review* 1, no. 1 (2014): 57-61, https://doi.org/10.1177/2053019613516

36. Cameron Harrington, "The Ends of the World: International Relations and the Anthropocene," *Millennium* 44, no. 3 (2016): 478-498, https://doi.org/10.1177/ 0305829816638745

37. John M. Meyer, *Political Nature: Environmentalism and the Interpretation of Western Thought* (MIT Press, 2001), 21.

38. Naomi Oreskes, "Science and Public Policy: What's Proof Got to Do with It?" *Environmental Science and Policy* 7, no. 5 (2004): 369–383, 381, https://doi.org/10.1016/j.envsci.2004.06.002

39. Naomi Oreskes, *Why Trust Science?* (Princeton University Press, 2019), 55.

40. Oreskes, *Why Trust Science?*

41. Myanna Lahsen, "We Cannot Afford Not to Perform Constructionist Studies of Mainstream Climate Science," in *Climate Science and Society: A Primer,* ed. Zeke Baker, Tamar Law, Mark Vardy and Stephen Zehr (Earthscan Routledge, 2024), 29–38.

42. Lahsen, "We Cannot Afford Not to Perform Constructionist Studies," 36.

43. Oreskes, *Why Trust Science?,* 55.

44. Naomi Oreskes, "The Scientific Consensus on Climate Change," *Science* 306, no. 5702 (2004): 1686, https://doi.org/10.1126/science.1103618

45. John Cook et al., "Quantifying the Consensus on Anthropogenic Global Warming in the Scientific Literature," *Environmental Research Letters* 8, no. 2 (2013): 024024, https://doi.org/10.1088/1748-9326/8/2/024024

46. James Powell, "Scientists Reach 100% Consensus on Anthropogenic Global Warming," *Bulletin of Science, Technology & Society* 37, no. 4 (2017): 183–184, https://doi.org/10.1177/0270467619886266

47. IPCC, *Climate Change 2021: The Physical Science Basis,* Contribution of Working Group I to the Sixth Assessment Report of the Intergovernmental Panel on Climate Change, ed. Valérie Masson-Delmotte et al. (Cambridge University Press, 2023), https://doi.org/10.1017/9781009157896

48. Keynyn Brysse et al., "Climate Change Prediction: Erring on the Side of Least Drama?," *Global Environmental Change* 23, no. 1 (2013): 327–337, https://doi.org/10.1016/j.gloenvcha.2012.10.008

49. Brysse et al., "Climate Change Prediction," 9.

50. "Prof. Jason Box," *The Tipping Points,* accessed October 29, 2024, https://www.thetippingpoints.com/scientists/prof-jason-box/

51. Jeff Goodell, "Greenland Melting," *Rolling Stone,* accessed October 29, 2024, https://www.rollingstone.com/interactive/feature-greenland-melting/

52. Michael Oppenheimer et al., *Discerning Experts: The Practices of Environmental Assessment for Environmental Policy* (University of Chicago Press, 2019), 219.

53. Quoted in John H. Richardson, "When the End of Civilization Is Your Day Job," *Esquire,* July 20, 2018, accessed October 30, 2024, https://www.esquire.com/news-politics/a36228/ballad-of-the-sad-climatologists-0815/

54. William Ripple et al., "World Scientists' Warning of a Climate Emergency," *BioScience* 70, no. 1 (2020): 8–12, https://doi.org/10.1093/biosci/biz088

55. "Making Peace with Nature," *UN Environment Programme* (UNEP), February 11, 2021, 13, 15, https://www.unep.org/resources/making-peace-nature

56. I have used "new environmental paradigm" for discussion of issues before 2000 and "new ecological paradigm" for discussion after that date.

57. Naomi Oreskes and Erik M. Conway, *Merchants of Doubt: How a Handful of*

Scientists Obscured the Truth on Issues from Tobacco Smoke to Climate Change (Bloomsbury Press, 2010).

Chapter 1

1. Suzanne Simard, *Finding the Mother Tree: Discovering the Wisdom of the Forest* (Knopf, 2021).

2. Riley E. Dunlap, "The New Environmental Paradigm Scale: From Marginality to Worldwide Use," *Journal of Environmental Education* 40, no.1 (2008): 3–18, doi: 10.3200/joee.40.1.3-18; John C. Pierce et al., "Culture, Politics and Mass Publics: Traditional and Modern Supporters of the New Environmental Paradigm in Japan and the United States," *Journal of Politics* 49, no. 1 (1987): 54–79, https://www.journals.uchicago.edu/doi/abs/10.2307/2131134; Larry Shetzer et al., "Business-Environment Attitudes and the New Environmental Paradigm," *Journal of Environmental Education* 22, no. 4 (1991): 14–21, https://doi.org/10.1080/00958964.1991.9943057; Paul Stern et al., "The New Ecological Paradigm in Social-Psychological Context," *Environment and Behavior* 27, no. 6 (1995): 723–743, https://doi.org/10.1177/00139165952760

3. According to Rokeach, "primitive beliefs" form the most central core of an individual's belief system and represent "basic truths about physical reality, social reality and the nature of the self." Milton Rokeach, *Beliefs, Attitudes and Values* (Jossey-Bass, 1968), 6.

4. Some researchers have challenged the construct reliability of the NEP Scale, arguing that it has been used "inter-changeably as a measure of beliefs, attitudes, values, and worldviews." Jennifer Bernstein, "(Dis)Agreement over What? The Challenge of Quantifying Environmental Worldviews," *Journal of Environmental Studies and Sciences* 10, no. 2 (2020): 169–177, 172, https://doi.org/10.1007/s13412-020-00593-x. Some scholars argue that the NEP Scale is unidimensional, whereas others argue that environmental attitudes are a multidimensional construct. Taciano L. Milfont and John Duckitt, "The Environmental Attitudes Inventory: A Valid and Reliable Measure to Assess the Structure of Environmental Attitudes," *Journal of Environmental Psychology* 30, no. 1 (2010): 80–94, 81, https://doi.org/10.1016/j.jenvp.2009.09.001

5. Dunlap, "The New Environmental Paradigm Scale."

6. C. Lundmark, "The New Ecological Paradigm Revisited: Anchoring the NEP Scale in Environmental Ethics," *Environmental Education Research* 13, no. 3 (2007): 329–347, https://doi.org/10.1080/13504620701430448

7. Annick Hedlund-de Witt, "Exploring Worldviews and Their Relationships to Sustainable Lifestyles: Towards a New Conceptual and Methodological Approach," *Ecological Economics* 84 (2012): 74–83, 75, https://doi.org/10.1016/j.ecolecon.2012.09.009

8. Bernstein, "(Dis)Agreement."

9. Bernstein, "(Dis)Agreement."

10. Bernstein, "(Dis)Agreement," 172.

11. Riley E. Dunlap, "Trends in Public Opinion Toward Environmental Issues:

1965-1990," in *American Environmentalism: The U.S. Environmental Movement, 1970-1990*, ed. Riley E. Dunlap and Angela G. Mertig (Taylor & Francis, 1992), 89-116. The 2000 revision of the original NEP Scale also was intended to add balance between pro- and anti-NEP statements and to eliminate outdated sexist terminology. The original NEP Scale also was criticized for its psychometric adequacy. Adam C. Davis and Mirella L. Stroink, "The Relationship Between Systems Thinking and the New Ecological Paradigm," *Systems Research and Behavioral Science* 33, no. 4 (2016): 575-586, https://onlinelibrary.wiley.com/doi/abs/10.1002/sres.2371

12. According to Dunlap et al., the five dimensions of the ecological worldview on the NEP Scale include the reality of limits to growth, anti-anthropocentrism, the fragility of nature's balance, rejection of exemptionalism, and the possibility of an eco-crisis. Riley E. Dunlap et al., "Measuring Endorsement of the New Ecological Paradigm: A Revised NEP Scale," *Journal of Social Issues* 56, no. 3 (2000): 425-442, 432, https://doi.org/10.1111/0022-4537.00176

13. C. Xiao et al., "Ecological Worldview as the Central Component of Environmental Concern: Clarifying the Role of the NEP," *Society & Natural Resources* 32, no. 1 (2019): 53-72, 54, https://doi.org/10.1080/08941920.2018.1501529

14. Adam C. Davis and Mirella L. Stroink, "The Relationship Between Systems Thinking and the New Ecological Paradigm," *Systems Research and Behavioral Science* 33, no. 4 (2016): 575-586, 583, https://doi.org/10.1002/sres.2371

15. John Muir Exhibit, "John Muir Misquoted: 'When One Tugs at a Single Thing in Nature, He Finds It Attached to the Rest of the World' but the Correct Quote Is Actually 'When We Try to Pick out Anything by Itself, We Find It Hitched to Everything Else in the Universe,'" *Sierra Club*, accessed October 29, 2024, https://vault.sierraclub.org/john_muir_exhibit/writings/misquotes.aspx

16. Simard, *Finding the Mother Tree.*

17. Simard, *Finding the Mother Tree.*

18. Davis and Stroink, "The Relationship Between Systems Thinking," 584.

19. Brian Walker and David Salt, *Resilience Thinking: Sustaining Ecosystems and People in a Changing World* (Island Press, 2006), 1.

20. Davis and Stroink, "The Relationship Between Systems Thinking," 584.

21. Davis and Stroink, "The Relationship Between Systems Thinking," 583-584.

22. Davis and Stroink, "The Relationship Between Systems Thinking," 578.

23. Anne Pender, "From Partial to Integrated Perspectives: How Understanding Worldviews Can Expand Our Capacity for Transformative Climate Governance," *Earth System Governance* 16 (2023): 100174, doi:10.1016/j.esg.2023.100174

24. Annick de Witt, "Climate Change and the Clash of Worldviews: An Exploration of How to Move Forward in a Polarized Debate," *Zygon: Journal of Religion and Science* 50, no. 4 (2015): 906-921, doi:10.1111/zygo.12226

25. Shannon M. McNeeley and Heather Lazrus, "The Cultural Theory of Risk for Climate Change Adaptation," *Weather, Climate, and Society* 6, no. 4 (2014): 506-519, doi:10.1175/WCAS-D-13-00027.1

26. Cited in Jon Kohl, "Talking Climate with Those Holding Different World-

views," *Yale Climate Connections*, June 4, 2021, accessed October 30, 2024, http://yaleclimateconnections.org/2021/06/talking-climate-with-those-holding-different-worldviews/

27. John M. Meyer, *Political Nature: Environmentalism and the Interpretation of Western Thought* (MIT Press, 2001).

28. Zachary A. Smith, *The Environmental Policy Paradox* (Routledge, 2018); Michael E. Kraft, *Environmental Policy and Politics* (Routledge, 2018).

29. Lester Milbrath, *Environmentalists: Vanguard for a New Society* (SUNY Press, 1984).

30. Riley E. Dunlap and Kent D. Van Liere, "Commitment to the Dominant Social Paradigm and Concern for Environmental Quality," *Social Science Quarterly* 65, no. 4 (1984): 1013-1028, 1014, accessed October 29, 2024, https://www.research gate.net/publication/260419263_Commitment_to_the_Dominant_Social_Para digm_and_Concern_for_Environmental_Quality

31. Stephen F. Cotgrove, *Catastrophe or Cornucopia: The Environment, Politics and the Future* (Wiley and Sons, 1982), 33.

32. Gifford Pinchot, *Breaking New Ground* (Harcourt Brace and Company, 1947).

33. Simard, *Finding the Mother Tree*.

34. Naomi Oreskes and Erik M. Conway, *The Big Myth: How American Business Taught Us to Loathe Government and Love the Free Market* (Bloomsbury Publishing, 2023).

35. Richard N. L. Andrews, "Environmental Regulation and Business 'Self-Regulation,'" *Policy Sciences* 31, no. 3 (1998): 177-197.

36. 42 U.S.C. §13101 et seq. (https://www.epa.gov/laws-regulations/summary -pollution-prevention-act)

37. Cited in Andrews, "Environmental Regulation," 177.

38. Jacqueline Vaughn and Hanna Cortner, "Using Parallel Strategies to Promote Change: Forest Policymaking Under George W. Bush," *Review of Policy Research* 21, no. 6 (2004): 767-782, 770, doi:10.1111/j.1541-1338.2004.00107.x

39. During the first Trump administration, the *New York Times* kept a running total of the suspended and revised environmental rules, bringing unusual media attention to environmental rulemaking. In June 2024, there was considerable media coverage of the regulatory arena after the Supreme Court overturned the Chevron Doctrine in *Chevron vs. NRDC* ("*Chevron U.S.A., Inc. v. Natural Resources Defense Council, Inc.*," 467 U.S. 837 [1984]). The Chevron Doctrine enabled agencies like the EPA to interpret ambiguous statutes when promulgating environmental rules. Overturning Chevron will make it more difficult for regulatory agencies such as the EPA, the Food and Drug Administration, and the Federal Aviation Administration to promulgate regulations to protect environmental health and public safety.

40. Judith A. Layzer, *Open for Business: Conservatives' Opposition to Environmental Regulation* (MIT Press, 2012), 5.

41. Cited in Walter A. Rosenbaum, *Environmental Politics and Policy* (Sage/CQ Press, 2020), 49.

42. Layzer, *Open for Business*, 4.

43. Daniel A. Mazmanian and Michael E. Kraft, *Toward Sustainable Communities: Transition and Transformations in Environmental Policy* (MIT Press, 2009).

44. Mazmanian and Kraft, *Toward Sustainable Communities*, 319–320.

45. "Special Report: Agenda 21 and How to Stop It," *American Policy Center*, accessed October 30, 2024, https://americanpolicy.org/files/booklet.pdf

46. "Agenda 21: The UN, Sustainability and Right-Wing Conspiracy Theory," *Southern Poverty Law Center*, April 1, 2014, accessed December 6, 2024, https://www.splcenter.org/20140331/agenda-21-un-sustainability-and-right-wing-conspiracy-theory

47. "Agenda 21: The UN, Sustainability and Right-Wing Conspiracy Theory."

48. "Agenda 21 Conspiracy Theories Spread by Extremists, Politicians Pose Real Dangers," *Southern Poverty Law Center*, April 13, 2014, accessed October 30, 2024, https://www.splcenter.org/news/2014/04/13/new-splc-report-agenda-21-conspiracy-theories-spread-extremists-politicians-pose-real

49. Rachelle Peterson and Peter Wood, "Sustainability: Higher Education's New Fundamentalism," *National Association of Scholars*, accessed October 29, 2024, https://www.nas.org/storage/app/media/images/documents/NAS-Sustainability-Digital.pdf

50. Mark C. J. Stoddart et al., "The Contours of Anti-Environmentalism: An Introduction to the *Handbook of Anti-Environmentalism*," in *Handbook of Anti-Environmentalism*, ed. David Tindall, Mark C. J. Stoddart, and Riley E. Dunlap (Edward Elgar, 2022), 2–22, 8.

51. Hazel Erskine, "The Polls: Pollution and Its Costs," *Public Opinion Quarterly* 36, no. 1 (1972): 120–135, 120, https://doi.org/10.1086/267984

52. Peter J. Jacques et al., "The Organisation of Denial: Conservative Think Tanks and Environmental Scepticism," *Environmental Politics* 17, no. 3 (2008): 349–385.

53. Thomas. S. Kuhn, *The Structure of Scientific Revolutions* (University of Chicago Press, 2012).

54. Hanna J. Cortner and Margaret A. Moote, *The Politics of Ecosystem Management* (Island Press, 1998), 39.

55. Naomi Oreskes, *Plate Tectonics: An Insider's History of the Modern Theory of the Earth* (CRC Press, 2003).

56. Cotgrove, *Catastrophe or Cornucopia*, 88.

57. Naomi Klein, *This Changes Everything: Capitalism vs. the Climate* (Simon & Schuster, 2014).

58. Dennis C. Pirages and Paul R. Ehrlich, *Ark II: Social Responses to Environmental Imperatives* (W. H. Freeman, 1974), 47.

Chapter 2

1. Diane Dumanoski, *The End of the Long Summer: Why We Must Remake Our Civilization to Survive on a Volatile Earth* (Three Rivers Press, 2009).

2. United Nations, the World Commission on Environment and Development, *Our Common Future* (Oxford University Press, 1987), 22.

3. Will Steffen et al., "The Anthropocene: Are Humans Now Overwhelming the Great Forces of Nature?" *Ambio* 36, no. 8 (2007): 614–621, https://www.jstor.org/stable/25547826

4. Bill McKibben, *Eaarth: Making a Life on a Tough New Planet* (St. Martin's Griffin, 2011).

5. Clive Hamilton, *Defiant Earth: The Fate of Humans in the Anthropocene* (Polity Press, 2017), 9–10.

6. Will Steffen et al., "The Emergence and Evolution of Earth System Science," *Nature Reviews Earth & Environment* 1, no. 1 (2020): 54–63, 54, https://doi.org/10.1038/s43017-019-0005-6

7. Hamilton, *Defiant Earth*, 12.

8. "Welcome to the Anthropocene," accessed October 30, 2024, https://www.anthropocene.info/great-acceleration.php

9. William F. Ruddiman et al., "Defining the Epoch We Live In," *Science* 348, no. 6230 (2015): 38–39, https://doi.org/10.1126/science.aaa7297

10. Emily Chung, "Canada's Crawford Lake Chosen as 'Golden Spike' to Mark Proposed New Epoch," *CBS News Science*, July 11, 2023, accessed October 30, 2024, https://www.cbc.ca/news/science/crawford-lake-anthropocene-1.6902999

11. Erle C. Ellis, "The Anthropocene Is Not an Epoch—but the Age of Humans Is Most Definitely Underway," *The Conversation*, March 5, 2024, accessed October 30, 2024, http://theconversation.com/the-anthropocene-is-not-an-epoch-but-the-age-of-humans-is-most-definitely-underway-224495

12. Philip Gibbard et al., "The Anthropocene as an Event, Not an Epoch," *Journal of Quaternary Science* 37, no. 3 (2022): 395–399, https://doi.org/10.1002/jqs.3416

13. Ellis, "The Anthropocene Is Not an Epoch," 204.

14. Steffen et al., "The Anthropocene: Are Humans Now Overwhelming," 614.

15. Christopher Flavelle, "Climate Change Could Cut World Economy by $23 Trillion in 2050, Insurance Giant Warns," *New York Times*, November 4, 2021, https://www.nytimes.com/2021/04/22/climate/climate-change-economy.html

16. David W. Orr, *Dangerous Years: Climate Change, the Long Emergency and the Way Forward* (Yale University Press, 2016), 37.

17. Mark Lynas, *The God Species: How the Planet Can Survive the Age of Humans* (HarperCollins, 2011), 8.

18. Michael Shellenberger and Ted Nordhaus, *Break Through: Why We Can't Leave Saving the Planet to Environmentalists* (Mariner Books, 2009), 135.

19. Bill McKibben, *The End of Nature* (Random House, 1989).

20. Julie Dunlap and Susan A. Cohen, *Coming of Age at the End of Nature: A Generation Faces Living on a Changed Planet* (Trinity University Press, 2016); Paul

Wapner, *Living Through the End of Nature: The Future of American Environmentalism* (MIT Press, 2010).

21. Mark A. Maslin and Simon L. Lewis, "Anthropocene: Earth System, Geological, Philosophical and Political Paradigm Shifts," *Anthropocene Review* 2, no. 2 (2015): 108-116, 109, https://doi.org/10.1177/2053019615588791

22. Susan Baker, "Novel Ecosystems and the Return of Nature in the Anthropocene," in *Rethinking the Environment for the Anthropocene: Political Theory and Socionatural Relations in the New Geological Epoch,* ed. Manuel Arias-Maldonado and Zev Trachtenberg (Routledge, 2019), 51-64.

23. Baker, "Novel Ecosystems."

24. Baker, "Novel Ecosystems.

25. Nathaniel Morse et al., "Novel Ecosystems in the Anthropocene: A Revision of the Novel Ecosystem Concept for Pragmatic Applications," *Ecology and Society* 19, no. 2 (2014): 12, http://dx.doi.org/10.5751/ES-06192-190212

26. Morse et al., "Novel Ecosystems."

27. Anne Fremaux, "The Return of Nature in the Capitalocene: A Critique of the Ecomodernist Version of the 'Good Anthropocene,'" in *Rethinking the Environment for the Anthropocene: Political Theory and Socionatural Relations in the New Geological Epoch*, ed. Manuel Arias-Maldonado and Zev Trachtenberg (Routledge, 2019), 19-36, 23.

28. Fremaux, "The Return of Nature," 26, 30.

29. Renée Cho, "How Close Are We to Climate Tipping Points?" *State of the Planet,* Columbia Climate School, November 11, 2021, accessed October 30, 2024, https://news.climate.columbia.edu/2021/11/11/how-close-are-we-to-climate-tipping-points/

30. William Ripple et al., "World Scientists' Warning of a Climate Emergency," *BioScience* 70, no. 1 (2020): 8-12, https://doi.org/10.1093/biosci/biz088.

31. National Research Council, *Abrupt Impacts of Climate Change: Anticipating Surprises* (National Academies Press, 2013), https://doi.org/10.17226/18373

32. Alexandra Witze, "Wally Broecker Divined How the Climate Could Suddenly Shift," *ScienceNews* March 29, 2022, accessed October 30, 2024, https://www.sciencenews.org/article/climate-change-wally-broecker-abrupt-ocean-conveyor-belt

33. National Research Council, *Abrupt Impacts,* vii.

34. Michael Mann and Tom Toles, *The Madhouse Effect: How Climate Change Denial Is Threatening Our Planet, Destroying Our Politics, and Driving Us Crazy* (Columbia University Press, 2016), 128.

35. Cited in Emily Pontecorvo, "Can the World Overshoot Its Climate Targets—and Fix It Later?," *Grist,* March 30, 2022, accessed October 30, 2024, https://grist.org/science/can-the-world-overshoot-its-climate-targets-and-then-fix-it-later/

36. Jerome R. Ravetz, "Post-Normal Science and the Complexity of Transitions Towards Sustainability," *Ecological Complexity* 3, no. 4 (2006): 275-284, https://doi.org/10.1016/j.ecocom.2007.02.001

37. Dumanoski, *End of the Long Summer,* 5.

38. Fremaux, "The Return of Nature," 28.

39. Peter de Menocal, "Wallace Smith Broecker (1931-2019)," *Nature* 568, no. 7750 (2019): 34-34, accessed October 30, 2024, https://www.nature.com/articles/d41586-019-00993-2

40. "1992 World Scientists' Warning to Humanity," Union of Concerned Scientists, July 16, 1992, accessed October 31, 2024, https://www.ucsusa.org/resources/1992-world-scientists-warning-humanity

41. IPCC, *Climate Change 2021: The Physical Science Basis,* Contribution of Working Group I to the Sixth Assessment Report of the Intergovernmental Panel on Climate Change, ed. Valérie Masson-Delmotte et al. (Cambridge University Press, 2023), https://doi.org/10.1017/9781009157896

42. Mathew Humphrey, "Democratic Legitimacy, Public Justification and Environmental Direct Action," *Political Studies* 54, no. 2 (2006): 310-327, https://doi.org/10.1111/j.1467-9248.2006.00602.x

43. Humphrey, "Democratic Legitimacy"; Steve Vanderheiden, "Eco-Terrorism or Justified Resistance? Radical Environmentalism and the 'War on Terror,'" *Politics and Society* 33, no. 3 (2005): 425-447, https://doi.org/10.1177/0032329205278462

44. George Hoberg, "Science, Politics, and U.S. Forest Service Law: The Battle over the Forest Service Planning Rule," *Natural Resources Journal* 44, no. 1 (2004): 1-27, 8, https://www.jstor.org/stable/24888909

45. Stephen R. Dovers, "Sustainability: Demands on Policy," *Journal of Public Policy* 16, no. 3 (1996): 303-318, https://www.jstor.org/stable/4007649

46. Clive Hamilton et al., *The Anthropocene and the Global Environmental Crisis: Rethinking Modernity in a New Epoch* (Routledge, 2015), 10.

47. Hamilton et al., *The Anthropocene and the Global Environmental Crisis,* 11.

48. Ewa Bińczyk, "The Most Unique Discussion of the 21st Century? The Debate on the Anthropocene Pictured in Seven Points," *Anthropocene Review* 6, no. 1-2 (2019): 3-18, 9, https://doi.org/10.1177/2053019619848215

49. Hamilton et al., *The Anthropocene and the Global Environmental Crisis,* 10.

50. As cited in *Merriam-Webster:*

> Thomas Jefferson said, "The earth belongs in usufruct to the living." He apparently understood that when you hold something in usufruct, you gain something of significant value, but only temporarily. The gains granted by usufruct can be clearly seen in the Latin phrase from which the word developed, *usus et fructus,* which means "use and enjoyment." Latin speakers condensed that phrase to *ususfructus,* the term English speakers used as the model for our modern word. *Usufruct* has been used as a noun for the legal right to use something since the mid-1600s. Any right granted by usufruct ends at a specific point, usually the death of the individual who holds it.

Accessed October 30, 2024, https://www.merriam-webster.com/dictionary/usufruct#note-1

51. Founders Online, "To James Madison from Thomas Jefferson, 6 September 1789," *National Archives,* accessed October 30, 2024, https://founders.archives.gov/documents/Madison/01-12-02-0248

Chapter 3

1. Michael P. Cohen, *The History of the Sierra Club 1892-1970* (Random House, 1988).

2. Eliot Porter and David Brower, eds., *The Place No One Knew: Glen Canyon on the Colorado* (Sierra Club, 1963).

3. Wallace Stegner, *This Is Dinosaur* (Knopf, 1955).

4. Cohen, *The History of the Sierra Club*.

5. Public Law 84-485; 70 Stat. 105.

6. Rachel Carson, *Silent Spring* (Houghton Mifflin, 1962).

7. "Our Story: How EDF Got Started," *Environmental Defense Fund*, accessed October 30, 2024, https://www.edf.org/about/our-history

8. "Fighting Polluters Since 1970," *Natural Resources Defense Council*, accessed October 30, 2024, https://nrdc50.org/

9. "Clean Air Act," 42 U.S.C. §7401 et seq. (1970).

10. "Clean Water Act," 33 U.S.C. §1251 et seq. (1972).

11. "Safe Drinking Water Act," Pub. L. No. 93-523, 88 Stat. 1660 (1974).

12. "Endangered Species Act of 1973," Pub L. No. 93-205, 87 Stat. 884, 16 U.S.C. §§ 1531-1544.

13. Christopher J. Bosso, *Environment, Inc.: From Grassroots to Beltway* (University Press of Kansas, 2005).

14. Philip Shabecoff, "Environmental Groups Told They Are Racists in Hiring," *New York Times*, February 1, 1990, https://www.nytimes.com/1990/02/01/us/environmental-groups-told-they-are-racists-in-hiring.html

15. "Environmental Justice Policy," *Sierra Club*, accessed October 30, 2024, https://www.sierraclub.org/policy/environmental-justice

16. "Environmental Justice Policy," *Sierra Club*.

17. "Environmental Justice Policy," *National Wildlife Federation*, accessed October 30, 2024, https://www.nwf.org/Our-Work/Environmental-Justice

18. "Environment, Equity & Justice Center," *National Resource Defense Council*, accessed October 30, 2024, https://www.nrdc.org/about/environment-equity-justice-center

19. "Climate Justice: Addressing Injustices for a Better Future," *Environmental Defense Fund*, accessed October 30, 2024, https://www.edf.org/climate-environmental-justice

20. "Climate Change," *National Resources Defense Council*, accessed October 30, 2024, https://www.nrdc.org/issues/climate-change#solutions

21. Carson, *Silent Spring*.

22. Hazel Erskine, "The Polls: Pollution and Its Costs," *Public Opinion Quarterly* 36, no. 1 (1972): 120-135, 120, https://doi.org/10.1086/267984

23. Aaron McCright et al., "Political Polarization on Support for Government Spending on Environmental Protection in the USA, 1974-2012," *Social Science Research* 48 (2014): 251-260, https://doi.org/10.1016/j.ssresearch.2014.06.008

24. "Can Environmentalism Become a Bipartisan Movement Again?" *PBS New-*

shour, April 25, 2016, http://www.pbs.org/newshour/bb/can-environmentalism -become-a-bipartisan-movement-again/

25. Frederick Buell, *From Apocalypse to Way of Life: Environmental Crisis in the American Century* (Routledge, 2003).

26. James Morton Turner, "'The Specter of Environmentalism': Wilderness, Environmental Politics, and the Evolution of the New Right," *Journal of American History* 96, no. 1 (2009): 123–148, https://www.jstor.org/stable/27694734

27. James Conaway, "James Watt, in the Right with the Lord," *Washington Post,* April 27, 1983, https://www.washingtonpost.com/archive/lifestyle/1983/04/27/ james-watt-in-the-right-with-the-lord/946026f4-3b79-4fcb-9a5f-082ca8150007/

28. Calvin E. Beisner, "The Competing World Views of Environmentalism and Christianity," *Cornwall Alliance,* accessed October 31, 2024, https://www.corn wallalliance.org/wp-content/uploads/2014/04/Competing-Worldviews-of-Envi ronmentalism-and-Christianity.pdf; see also Joel Garreau, "Environmentalism as Religion," *New Atlantis,* Summer 2010, accessed October 31, 2024, https://www. thenewatlantis.com/publications/environmentalism-as-religion

29. Michael E. Kraft and Norman J. Vig, "Environmental Policy in the Reagan Presidency," *Political Science Quarterly* 99, no. 3 (1984): 415–439, https://doi.org/10 .2307/2149941

30. Kraft and Vig, "Environmental Policy in the Reagan Presidency."

31. Alyssa Serrani, "James G. Watt Quotes That Speak to What He Believed In," *Your Dictionary,* accessed October 31, 2024, https://www.yourdictionary.com/ articles/james-g-watt-quotes-beliefs

32. Jonathan S. Coley and Jessica Schachle, "Growing the Green Giant: Ecological Threats, Political Threats, and U.S. Membership in Sierra Club,1892–Present," *Social Sciences* 10, no. 6 (2021): 189, https://doi.org/10.3390/socsci10060189

33. Jeffrey M. Jones, "Four in 10 Americans Say They Are Environmentalists," *Gallup,* April 21, 2021, accessed October 31, 2024, https://news.gallup.com/poll/ 348227/one-four-americans-say-environmentalists.aspx#:~:text=When%20 Gallup%20first%20asked%20this,three%20polls%20conducted%20since%20 2016

34. "Environment," *Gallup,* accessed October 31, 2024, https://news.gallup.com /poll/1615/Environment.aspx

35. "II. Seeing the Forest for the Trees: Placing Washington's Forests in Historical Context," *Center for the Study of the Pacific Northwest,* accessed October 31, 2024, https://www.washington.edu/uwired/outreach/cspn/Website/Classroom %20Materials/Curriculum%20Packets/Evergreen%20State/Section%20II.html

36. Harley Rustad, "Big Lonely Doug," *The Walrus,* October 2016, accessed October 31, 2024, https://thewalrus.ca/big-lonely-doug/

37. "Quinault Big Sitka Spruce Tree," *National Park Service,* accessed October 31, 2024, https://www.nps.gov/places/000/quinault-big-sitka-spruce-tree.htm #:~:text=From%20the%20resort%20which%20manages,inches%20and%20191% 20feet%20tall.%22

38. "The Forest and the Trees," *Earthjustice*, May 12, 2002, accessed October 31, 2024, https://earthjustice.org/feature/the-forest-and-the-trees#:~:text=By%20 the%201980s%2C%20the%2017,forest%20once%20started%20to%20unravel

39. "National Forest Management Act," Pub. L. No. 94-588, 90 Stat. 2949 (1976); "Endangered Species Act."

40. James Wolcott, "Rush to Judgment," *Vanity Fair*, May 2007, https://www .vanityfair.com/news/2007/05/wolcott200705

41. "Multiple Use Sustained Yield Act of 1960," 16 U.S.C. § 528.

42. "General Mining Act of 1872," U.S.C. § 22 et seq.

43. Mark C. J. Stoddart et al., "The Contours of Anti-Environmentalism: An Introduction to the *Handbook of Anti-Environmentalism*," in *Handbook of Anti-Environmentalism*, ed. David Tindall, Mark C. J. Stoddart, and Riley E. Dunlap (Edward Elgar, 2022), 2-22, 6.

44. Dan Zak, "A Green New Deal Ignites an Old Red Scare," *Washington Post*, May 8, 2019, https://www.washingtonpost.com/lifestyle/style/a-green-new-deal-ignites -an-old-red-scare/2019/05/07/6f65be80-62df-11e9-9412-daf3d2e67c6d_story.html

45. Timothy Egan, "Fund-Raisers Tap Anti-Environmentalism," *New York Times*, December 19, 1991, https://www.nytimes.com/1991/12/19/us/fund-raisers -tap-anti-environmentalism.html

46. "Center for the Defense of Free Enterprise (CDFE)," *DeSmog*, accessed October 31, 2024, https://www.desmog.com/center-defense-free-enterprise/

47. Dixie Lee Ray, "Science and the Environment," *Religion and Liberty* 2, no. 7 (2010), accessed October 31, 2024, https://www.acton.org/pub/religion-liberty/ volume-2-number-7/science-and-environment

48. Paul Krugman, "Why Republicans Turned Against the Environment," *New York Times*, August 15, 2022, https://www.nytimes.com/2022/08/15/opinion/ republicans-environment-climate.html

49. Executive Order 14008, 86 Fed. Reg. 2021 (February 1, 2021), https://www .govinfo.gov/content/pkg/FR-2021-02-01/pdf/2021-02177.pdf

50. "Justice40 Initiative," *Office of Energy Justice and Equity*, accessed October 31, 2024, https://www.energy.gov/justice/justice40-initiative

51. Brian Kennedy et al., "How Americans See Biden's Climate Policies," *Pew Research Center*, June 28, 2023, accessed October 31, 2024, https://www.pewre search.org/science/2023/06/28/2-how-americans-see-bidens-climate-policies/

52. Buell, *From Apocalypse to Way of Life*, 18.

53. Buell, *From Apocalypse to Way of Life*, 21.

54. Nadia Y. Bashir et al., "The Ironic Impact of Activists: Negative Stereotypes Reduce Social Change Influence," *European Journal of Social Psychology* 43, no. 7 (2013): 614-626, https://doi.org/10.1002/ejsp.1983

55. Anna Klas et al., "'Not All Environmentalists Are Like That . . .': Unpacking the Negative and Positive Beliefs and Perceptions of Environmentalists," *Environmental Communication* 13, no. 7 (2019): 879-893, https://doi.org/10.1080/17524032 .2018.1488755

56. Michael Shellenberger, *Apocalypse Never: Why Environmental Alarmism Hurts Us All* (Harper, 2020); Michael Shellenberger, "On Behalf of Environmentalists, I Apologize for the Climate Scare," *Environmental Progress*, June 29, 2020, accessed October 31, 2024, https://environmentalprogress.org/big-news/2020/6/29/on-behalf-of-environmentalists-i-apologize-for-the-climate-scare

57. Amy Westervelt, *Drilled*, Season 7, "Redefining Environmentalists," *Critical Frequency*, January 14, 2022, 15 min., 9 sec, accessed October 31, 2024, https://www.spreaker.com/episode/redefining-environmentalists--48289859

58. Joel Connelly, "Alaska's Don Young—Loose Lips of Congress' Longest Serving Republican," *SeattlePI*, March 6, 2019, accessed October 31, 2024, https://www.seattlepi.com/local/politics/article/Alaska-s-Don-Young-loose-lips-of-Congress-13668873.php

59. Suzanne Goldenberg, "Rush Limbaugh Goes the Extra Mile in Rant About New York Times Reporter," *The Guardian*, October 21, 2009, https://www.theguardian.com/environment/2009/oct/21/rush-limbaugh-andy-revkin

60. Shellenberger, *Apocalypse Never*.

61. James Wanliss, *Resisting the Green Dragon; Dominion, Not Death* (Cornwall Alliance, 2011).

62. Peter J. Jacques et al., "The Organisation of Denial: Conservative Think Tanks and Environmental Scepticism," *Environmental Politics* 17, no. 3 (2008): 349–385.

63. John M. Meyer, "Interpreting Nature and Politics in the History of Western Thought: The Environmentalist Challenge," *Environmental Politics* 8, no. 2 (1999): 1–23, 2, https://doi.org/10.1080/09644019908414459

64. D. E. Morrison and Riley E. Dunlap, "Environmentalism and Elitism: A Conceptual and Empirical Analysis," *Environmental Management* 10, no. 5 (1986): 581–589, https://doi.org/10.1007/BF01866762

65. William Tucker, ""Environmentalism and the Leisure Class," *Harper's Magazine* 225, no. 1531 (December 1, 1977), https://harpers.org/archive/1977/12/environmentalism-and-the-leisure-class/

66. Walter A. Rosenbaum, *Environmental Politics and Policy* (Sage/CQ Press, 2020).

67. Bosso, *Environment, Inc.*, 6.

68. Jones, "Four in 10 Americans."

69. Chris Benderev, "Millennials: We Help the Earth But Don't Call Us Environmentalists," *NPR*, October 11, 2014, accessed October 31, 2024, https://www.npr.org/2014/10/11/355163205/millennials-well-help-the-planet-but-dont-call-us-environmentalists

70. "Are Americans Concerned About Global Warming?" *Gallup*, October 5, 2021, accessed October 31, 2024, https://news.gallup.com/poll/355427/americans-concerned-global-warming.aspx

71. Anthony Leiserowitz et al., "Dramatic Increase in Public Beliefs and Worries About Climate Change," *Yale Program on Climate Change Communication*, September 7, 2021, accessed October 31, 2024, https://climatecommunication.yale

.edu/publications/dramatic-increase-in-public-beliefs-and-worries-about-climate
-change/

72. Alec Tyson and Brian Kennedy, "How Americans View Future Harms from
Climate Change in Their Community and Around the U.S.," *Pew Research Center*
(blog), October 25, 2023, accessed October 30, 2024, https://www.pewresearch.
org/science/2023/10/25/how-americans-view-future-harms-from-climate
-change-in-their-community-and-around-the-u-s/

73. Benderev, "Millennials: We Help the Earth."

74. Geoffrey Supran and Naomi Oreskes, "Rhetoric and Frame Analysis of ExxonMobil's Climate Change Communications," *One Earth* 4, no. 5 (2021), doi:10
.1016/j.oneear.2021.04.014

75 Earth Works Group, *Fifty Simple Things You Can Do to Save the Earth* (Turtle-
back, 1990).

76. Klas et al., "'Not All Environmentalists Are Like That.'"

77. Bashir et al., "The Ironic Impact of Activists," 620.

78. Frank Newport, "Update: Partisan Gaps Expand Most on Government
Power, Climate," *Gallup*, August 7, 2023, accessed October 30, 2024, https://news
.gallup.com/poll/509129/update-partisan-gaps-expand-government-power-cli
mate.aspx

79. Klas et al., "'Not All Environmentalists Are Like That,'" 889.

80. Bashir et al., "The Ironic Impact of Activists," 620.

81. *Dobbs v. Jackson Women's Health Organization*, 597 U.S. at 215 (2022).

82. Jon Devine, "Dirty Water Actors in Congress," *National Resources Defense
Council* (expert blog), February 16, 2011, accessed October 31, 2024, https://www
.nrdc.org/bio/jon-devine/dirty-water-actors-congress

83. National Environmental Scorecard, "1996 Scorecard Vote, Anti-Environment
Riders I, House Roll Call Vote 5," *League of Conservation Voters*, accessed October 31,
2024, https://scorecard.lcv.org/roll-call-vote/1996-5-anti-environment-riders-i

84. "Riders from Hell," *Washington Post*, October 24, 1995, https://www.wash
ingtonpost.com/archive/opinions/1995/10/25/riders-from-hell/35d8f806-928e
-4de8-a114-66668580b7e3/

85. Don Gonyea, "The Longest Government Shutdown in History, No Longer—
How 1995 Changed Everything," *NPR*, January 12, 2019, accessed October 31,
2024, https://www.npr.org/2019/01/12/683304824/the-longest-government-shut
down-in-history-no-longer-how-1995-changed-everything

86. *Loper Bright Enterprises, et al. v. Gina Raimondo, Secretary of Commerce, et
al.Relentless, Inc. et al. v. Department of Commerce, et al.*, 603 U.S. ___ (2024), 144
S. Ct. 2244.

87. Maxine Joselow, "Red States Are Blocking Blue Cities from Setting Climate
Policies," *Washington Post*, June 13, 2023, https://www.washingtonpost.com/poli
tics/2023/06/13/red-states-are-blocking-blue-cities-setting-climate-policies/

88. Natalie Delgadillo, "State and Local Officials Clash over Plastic Bag Bans,"
Governing, January 29, 2018, accessed October 31, 2024, https://www.governing
.com/archive/stl-plastic-bag-bans.html

89. Jones, "Four in 10 Americans."

90. "National Survey: American Voters View Conservation as Patriotic," *We-ConservePA*, accessed October 31, 2024, https://conservationtools.org/library_items/1205

Chapter 4

1. Hanna J. Cortner and Margaret Ann Moote, *The Politics of Ecosystem Management* (Island Press, 1998), 39.

2. Aaron M. McCright and Riley E. Dunlap, "Anti-Reflexivity," *Theory, Culture & Society* 27, no. 2-3 (2010): 100-133, https://doi.org/10.1177/0263276409356001

3. Naomi Oreskes and Erik M. Conway, *Merchants of Doubt: How a Handful of Scientists Obscured the Truth on Issues from Tobacco Smoke to Climate Change* (Bloomsbury Press, 2010).

4. "The Advancement of Sound Science Coalition (TASSC)," *DeSmog,* accessed November 17, 2024, https://www.desmog.com/advancement-sound-science-coalition/

5. Cited in Oliver Burkeman, "Memo Exposes Bush's New Green Strategy," *The Guardian,* March 3, 2003, https://www.theguardian.com/environment/2003/mar/04/usnews.climatechange

6. "H.R. 4840—Sound Science for Endangered Species Act Planning Act of 2002," https://www.congress.gov/bill/107th-congress/house-bill/4840

7. Peter J. Jacques et al., "The Organisation of Denial: Conservative Think Tanks and Environmental Scepticism," *Environmental Politics* 17, no. 3 (2008): 349-385.

8. Paul R. Ehrlich and Anne H. Ehrlich, *Betrayal of Science and Reason: How Anti-Environmental Rhetoric Threatens Our Future* (Island Press, 1996).

9. Jacques et al., "The Organisation of Denial."

10. Bjørn Lomborg, *The Skeptical Environmentalist: Measuring the Real State of the World* (Cambridge University Press, 2001).

11. "Doomsday Postponed," *Economist,* September 6, 2001, https://www.economist.com/books-and-arts/2001/09/06/doomsday-postponed

12. Ronald Bailey, "Why All Those Dire Predictions Have No Future," *Wall Street Journal,* October 2, 2001, https://www.wsj.com/articles/SB1001976791758229400

13. Neela Banerjee et al., "Exxon: The Road Not Taken," *Inside Climate News,* accessed October 31, 2024, https://insideclimatenews.org/book/exxon-the-road-not-taken/

14. Banerjee et al., "Exxon."

15. "The Greenhouse Effect," Greenhouse Effect Working Group, Shell International Health, Safety and Environmental Division, Environmental Affairs, April 1986, accessed October 31, 2024, https://www.documentcloud.org/documents/4411090-Document3.html#document/p4/a415539

16. Exxon merged with Mobil in 1999, forming ExxonMobil. Unless stated otherwise, I follow Supran and Oreskes (2017) in referring to ExxonMobil Corporation, Exxon Corporation, and Mobil Oil Corporation as ExxonMobil.

17. Banerjee et al., "Exxon."

18. Philip Shabecoff, "Global Warming Has Begun, Expert Tells Senate," *New York Times*, June 24, 1988, https://www.nytimes.com/1988/06/24/us/global-war ming-has-begun-expert-tells-senate.html

19. Ken Cohen, "When It Comes to Climate Change, Read the Documents," ExxonMobil's Perspectives archive, October 21, 2015, accessed October 31, 2024, https://perma.cc/533R-8PKY

20. Geoffrey Supran and Naomi Oreskes, "Assessing ExxonMobil's Climate Change Communications (1977–2014)," *Environmental Research Letters* 12 (2017): 084019, https://doi.org/10.1088/1748-9326/aa815f

21. C. Brown and H. Waltzer, "Every Thursday: Advertorials by Mobil Oil on the Op-Ed Page of *The New York Times*," *Public Relations Review* 31, no. 2 (2005): 197–208, https://doi.org/10.1016/j.pubrev.2005.02.019

22. Geoffrey Supran and Naomi Oreskes, "The Forgotten Oil Ads That Told Us Climate Change Was Nothing," *E = mc2: Energy Matters to Climate Change*, accessed November 1, 2024, https://www.e-mc2.gr/el/news/forgotten-oil-ads-told-us-cli mate-change-was-nothing

23. "Utilities Knew: Documenting Electric Utilities' Early Knowledge and Ongoing Deception on Climate Change," *Energy and Policy Institute*, accessed November 1, 2024, https://www.energyandpolicy.org/utilities-knew-about-climate-change/

24. Kyoto Protocol to the United Nations Framework Convention on Climate Change, 1997.

25. "Information Council for the Environment Documents," *Union of Concerned Scientists*, accessed November 1, 2024, https://www.ucsusa.org/sites/default/files/attach/2015/07/Climate-Deception-Dossier-5_ICE.pdf

26. "1988 American Petroleum Institute Global Climate Science Communications Team Action Plan," *ClimateFiles*, accessed November 1, 2024, http://www.cli matefiles.com/trade-group/american-petroleum-institute/1998-global-climate -science-communications-team-action-plan/

27. "1988 American Petroleum Institute."

28. "1988 American Petroleum Institute."

29. Dimitrios Gounaridis and Joshua P. Newell, "The Social Anatomy of Climate Change Denial in the United States," *Scientific Reports* 14, no. 2097 (2024), https://doi.org/10.1038/s41598-023-50591-6

30. Benjamin Franta, "Weaponizing Economics: Big Oil, Economic Consultants, and Climate Policy Delay," *Environmental Politics* 31, no. 4 (2021): 1-21.

31. Robert J. Brulle and Carter Werthman, "The Role of Public Relations Firms in Climate Change Politics," *Climatic Change* 169, no. 8 (2021), http://link.springer .com/10.1007/s10584-021-03244-4

32. Cited in Burkeman, "Memo Exposes Bush's New Green Strategy." See also https://www.sourcewatch.org/images/4/45/LuntzResearch.Memo.pdf, accessed November 1, 2024.

33. Cited in Burkeman, "Memo Exposes Bush's New Green Strategy."

34. Marlo Lewis Jr., "A Citizen's Guide to Climate Change," *Competitive Enter-

prise Institute, June 11, 2019, accessed November 1, 2024, https://cei.org/studies/a-citizens-guide-to-climate-change/

35. Patrick J. Michaels and Paul C. Knappenberger, *Lukewarming: The New Climate Science That Changes Everything* (Cato Institute, 2016).

36. Craig Idso et al., *Why Scientists Disagree About Global Warming: The NIPCC Report on Scientific Consensus* (Heartland Institute, 2018).

37. PragerU, accessed November 1, 2024, https://www.prageru.com/video/the-real-climate-crisis

38. Alex Epstein, *The Moral Case for Fossil Fuels* (Portfolio, 2014); Alex Epstein, *Fossil Future: Why Global Human Flourishing Requires More Oil, Coal, and Natural Gas—Not Less* (Portfolio, 2022).

39. Riley E. Dunlap et al., "The Political Divide on Climate Change: Partisan Polarization Widens in the U.S. Environment," *Science and Policy for Sustainable Development* 58, no. 5 (2016): 4–23, 15, https://doi.org/10.1080/00139157.2016.1208995

40. Andrew J. Hoffman, *How Culture Shapes the Climate Change Belief* (Stanford University Press, 2015), 1.

41. Albert Gore, *Earth in the Balance: Ecology and the Human Spirit* (Houghton Mifflin, 1992).

42. "Planet Gore," *National Review,* accessed November 1, 2024, https://www.nationalreview.com/planet-gore/

43. Glenn Beck and Kevin Balfe, *An Inconvenient Book: Real Solutions to the World's Biggest Problems* (Threshold Editions, 2007); Katie Machol Simon, "Glenn Beck's Environmental Greatest Hits," *Tampa Bay Creative Loafing,* April 7, 2011, accessed November 1, 2024, https://www.cltampa.com/news/glenn-becks-environmental-greatest-hits-video-12268945

44. Mark J. Perry, "Politics Disguised as Science: A 12-Point Checklist for When to Doubt a Scientific 'Consensus,'" *American Enterprise Institute* (blog), October 5, 2019, accessed November 1, 2024, https://www.aei.org/carpe-diem/politics-disguised-as-science-a-12-point-checklist-for-when-to-doubt-a-scientific-consensus/

45. Lisa Friedman, "What Is the Green New Deal? A Climate Proposal, Explained," *New York Times,* February 21, 2019, https://www.nytimes.com/2019/02/21/climate/green-new-deal-questions-answers.html

46. Ted Macdonald, "Fox News Discussed the Green New Deal More Often Than CNN and MSNBC Combined," *Media Matters for America,* April 9, 2019, accessed November 1, 2024, https://www.mediamatters.org/fox-news/fox-news-discussed-green-new-deal-more-often-cnn-and-msnbc-combined?redirect_source=/blog/2019/04/09/Fox-News-discussed-the-Green-New-Deal-more-often-than-CNN-and-MSNBC-combined/223383; Abel Gustafson et al., "The Development of Partisan Polarization over the Green New Deal," *Nature Climate Change* 9, no. 12 (2019): 940–944, https://doi.org/10.1038/s41558-019-0621-7

47. Bjørn Lomborg, *Cool It: The Skeptical Environmentalist's Guide to Global Warming* (Knopf, 2007); Bjorn Lomborg, *False Alarm: How Climate Change Panic Costs Us Trillions, Hurts the Poor, and Fails to Fix the Planet* (Basic Books, 2024).

48. George Will, "Opinion: With a Closer Look, Certainty About the 'Existen-

tial' Climate Threat Melts Away," *Washington Post,* August 11, 2021, https://www
.washingtonpost.com/opinions/2021/08/11/with-closer-look-certainty-about-ex
istential-climate-threat-melts-away/

49. Michael Mann, *The New Climate War* (PublicAffairs, 2022), 45.

50. Geoffrey Supran and Naomi Oreskes, "Rhetoric and Frame Analysis of ExxonMobil's Climate Change Communications," *One Earth* 4, no. 5 (2021), doi:10
.1016/j.oneear.2021.04.014

51. Rebecca Solnit, "Big Oil Coined 'Carbon Footprints' to Blame Us for Their
Greed. Keep Them on the Hook," *The Guardian,* August 23, 2021, https://www.the
guardian.com/commentisfree/2021/aug/23/big-oil-coined-carbon-footprints-to
-blame-us-for-their-greed-keep-them-on-the-hook

52. Elliott Hyman, "Who's Really Responsible for Climate Change?" *Harvard
Political Review,* January 2, 2020, accessed November 1, 2024, https://harvardpoli
tics.com/climate-change-responsibility/

53. Ian Schwartz, "Marc Morano: We Will Go from COVID Lockdowns to 'Climate Lockdowns' Under Biden," *Real Clear Politics,* December 21, 2020, video, 5
min., 3 sec., accessed November 1, 2024, https://www.realclearpolitics.com/video
/2020/12/21/marc_morano_we_will_go_from_covid_lockdowns_to_climate_
lockdowns_under_biden.html

54. Peter Sinclair, "'It's Almost Like a Cult.' Activists Shout Down Rural Renewable Energy Projects," *Yale Climate Connections,* February 28, 2023, accessed
November 1, 2024, http://yaleclimateconnections.org/2023/02/its-almost-like-a
-cult-activists-shout-down-rural-renewable-energy-projects/

55. Emma Frances Bloomfield and Denise Tillery, "The Circulation of Climate
Change Denial Online: Rhetorical and Networking Strategies on Facebook," *Environmental Communication* 13, no. 1 (2019): 23-34, 24, https://doi.org/10.1080/1752
4032.2018.1527378

56. Hiroko Tabuchi, "How Climate Change Deniers Rise to the Top in Google
Searches," *New York Times,* December 29, 2017, https://www.nytimes.com/2017/
12/29/climate/google-search-climate-change.html

57. Dawn Stover, "Why Facebook, YouTube, and Twitter Are Bad for the Climate," *Bulletin of the Atomic Scientists,* August 22, 2019, https://thebulletin.org/
2019/08/why-facebook-youtube-and-twitter-are-bad-for-the-climate/

58. Joachim Allgaier, "Science and Environmental Communication on You-
Tube: Strategically Distorted Communications in Online Videos on Climate
Change and Climate Engineering," *Frontiers in Communication,* 4 (2019), https://
doi.org/10.3389/fcomm.2019.00036

59. "The Toxic Ten: How Ten Fringe Publishers Fuel 69% of Digital Climate
Change Denial," *Center for Countering Digital Hate,* accessed November 1, 2024,
https://www.counterhate.com/toxicten

60. Cat Zakrzewski, "Facebook Whistleblower Alleges Executives Misled Investors About Climate, Covid Hoaxes in New SEC Complaint," *Washington Post,*
February 18, 2022, https://www.washingtonpost.com/technology/2022/02/18/
whistleblower-facebook-sec-climate-change/

61. Nick Robins-Early, "Twitter Ranks Worst in Climate Change Misinformation Report," *The Guardian,* September 20, 2023, https://www.theguardian.com/technology/2023/sep/20/twitter-x-musk-climate-misinformation-social-platforms

62. Scott Waldman, "Denial Expands on Facebook as Scientists Face Restrictions," *Politico,* July 6, 2020, https://subscriber.politicopro.com/article/eenews/1063511857

63. Rachel Aviv, "A Valuable Reputation," *New Yorker,* February 2, 2014, https://www.newyorker.com/magazine/2014/02/10/a-valuable-reputation

64. "Silencing the Scientist: Tyrone Hayes on Being Targeted by Herbicide Firm Syngenta," *Democracy Now!,* February 21, 2014, video, 31 min., 29 sec.; 49 min., 55 sec., accessed November 1, 2024, http://www.democracynow.org/2014/2/21/silencing_the_scientist_tyrone_hayes_on

65. See Michael E. Mann, *The Hockey Stick and the Climate Wars: Dispatches from the Front Lines* (Columbia University Press, 2012.)

66. Michael E. Mann, "I'm a Scientist Who Has Gotten Death Threats. I Fear What May Happen Under Trump," *Washington Post,* December 16, 2016, https://www.washingtonpost.com/opinions/this-is-what-the-coming-attack-on-climate-science-could-look-like/2016/12/16/e015cc24-bd8c-11e6-94ac-3d324840106c_story.html

67. *Cuccinelli v. Rector & Visitors of the University of Virginia, 283 Va. 420.*

68. Bradley A. Benbrook et al., "Competitive Enterprise Institute and National Review v. Michael E. Mann," *Cato Institute,* July 8, 2019, accessed November 1, 2024, https://www.cato.org/legal-briefs/competitive-enterprise-institute-national-review-v-michael-e-mann

69. Delger Erdenesanaa, "Michael Mann, a Leading Climate Scientist, Wins His Defamation Suit," *New York Times,* February 8, 2024, https://www.nytimes.com/2024/02/08/climate/michael-mann-defamation-lawsuit.html

70. Anthony A. Leiserowitz et al., "Climategate, Public Opinion, and the Loss of Trust," *American Behavioral Scientist* 57, no. 6 (2013): 818–837, https://doi.org/10.1177/0002764212458272

71. "Energy & Environment Legal Institute (E&E Legal)," *DeSmog,* accessed November 17, 2024, https://www.desmog.com/energy-environment-legal-institute/

72. "Silencing Scientists Tracker," *Columbia Climate School,* accessed December 30, 2024, https://climate.law.columbia.edu/Silencing-Science-Tracker

73. IPBES, "Summary for Policymakers of the Global Assessment Report on Biodiversity and Ecosystem Services of the Intergovernmental Science-Policy Platform on Biodiversity and Ecosystem Services," ed. S. Díaz et al., IPBES secretariat, Bonn, Germany, 2019, accessed November 1, 2024, https://files.ipbes.net/ipbes-web-prod-public-files/inline/files/ipbes_global_assessment_report_summary_for_policymakers.pdf

74. Julia Janicki, "The Collapse of Insects," *Reuters,* December 6, 2022, accessed November 1, 2024, https://graphics.reuters.com/GLOBAL-ENVIRONMENT/INSECT-APOCALYPSE/egpbykdxjvq/

75. Victor Anderson, "Biodiversity Loss Has Finally Got Political—and This

Means New Thinking on the Left and the Right," *The Conversation*, May 20, 2019, accessed November 17, 2024, http://theconversation.com/biodiversity-loss-has -finally-got-political-and-this-means-new-thinking-on-the-left-and-the-right-116 910

76. John R. Platt, "Rise of the Extinction Deniers," *Scientific American Blog Network*, June 22, 2019, https://blogs.scientificamerican.com/extinction-countdown/ rise-of-the-extinction-deniers/

77. Jimmy Tobias, "Republicans Aren't Just Climate Deniers. They Deny the Extinction Crisis, Too," *The Guardian*, May 23, 2019, https://www.theguardian. com/commentisfree/2019/may/23/republicans-arent-just-climate-deniers-they -deny-the-extinction-crisis-too

78. E. O. Wilson, *The Diversity of Life* (Belknap Press, 1992).

79. Julian L. Simon and Aaron Wildavsky, "Assessing the Empirical Basis of the 'Biodiversity Crisis,'" *Competitive Enterprise Institute*, May 1993, accessed November 1, 2024, https://cei.org/sites/default/files/Julian%20L%20%20Simon%20and %20Aaron%20Wildavsky%20-%20Assessing%20the%20Empirical%20Basis% 20of%20the%20Biodiversity%20Crisis.pdf

80. Ehrlich and Ehrlich, *Betrayal of Science and Reason*.

81. Platt, "Rise of the Extinction Deniers."

82. Alexander C. Lees et al., "Biodiversity Scientists Must Fight the Creeping Rise of Extinction Denial," *Nature Ecology & Evolution* 4, no. 11 (2020): 1440–1443, https://doi.org/10.1038/s41559-020-01285-z

83. "New UN Report on Looming Mass Extinctions Exposed in US House Testimony," *Heartland Institute*, May 28, 2019, accessed November 1, 2024, https:// heartland.org/opinion/new-un-report-on-looming-mass-extinctions-exposed-in -us-house-testimony/

84. Patrick Moore, *Fake Invisible Catastrophes and Threats of Doom* (self-published, 2024).

85. Marc Morano, "Greenpeace Co-Founder Dr. Patrick Moore's testimony to Congress on UN Species Report: UN Is Using 'Extinction as a Fear Tactic to Scare the Public into Compliance,'" *Climate Depot*, May 22, 2019, accessed November 1, 2024, https://www.climatedepot.com/2019/05/22/greenpeace-co-founder-dr-pat rick-moore-testimony-to-congress-on-un-species-report-un-is-using-extinction -as-a-fear-tactic-to-scare-the-public-into-compliance/

Chapter 5

1. The Clean Air Act of 1990 was championed by President George H. W. Bush. On the heels of the overwhelming public rejection of Reagan's anti-environmental policies, Bush distinguished himself as a presidential candidate by focusing on the environment.

2. Naomi Oreskes, "To Understand How Science Denial Works, Look to History," *Scientific American*, December 1, 2020, accessed November 4, 2024, https:// www.scientificamerican.com/article/to-understand-how-science-denial-works -look-to-history/

3. Naomi Oreskes, *Why Trust Science?* (Princeton University Press, 2019).

4. Oreskes, *Why Trust Science?*, 49.

5. Oreskes, *Why Trust Science?*, 59.

6. Hoyoon Jung, "The Evolution of Social Constructivism in Political Science: Past to Present," *Sage Open* 2, no. 1 (2019), https://doi.org/10.1177/2158244019832703

7. J. Gayon, "Realism and Biological Knowledge," in *Knowledge and the World: Challenges Beyond the Science Wars*, ed. Martin Carrier, Johannes Roggenhofer, Günter Küppers, and Philippe Blanchard (Springer, 2004), https://doi.org/10.1007 /978-3-662-08129-7_8

8. Elizabeth Ann R. Bird, "The Social Construction of Nature: Theoretical Approaches to the History of Environmental Problems," *Environmental History Review* 11, no. 4 (1987): 255-264, 255, https://doi.org/10.2307/3984134

9. S. Barry Barnes, "On Social Constructivist Accounts of the Natural Sciences," in *Knowledge and the World: Challenges Beyond the Science Wars*, ed. Martin Carrier, Johannes Roggenhofer, Günter Küppers, and Philippe Blanchard (Springer, 2004), 105-136, https://doi.org/10.1007/978-3-662-08129-7_5

10. Sven Ove Hansson, "Social Constructionism and Climate Science Denial," *European Journal for Philosophy of Science* 10, no. 37 (2020), https://doi.org/10.1007 /s13194-020-00305-w

11. H. M. Collins, "Stages in the Empirical Programme of Relativism," *Social Studies of Science* 11, no. 1 (1981): 3-10, 3, https://doi.org/10.1177/030631278101100101

12. Christopher Butler, *Postmodernism: A Very Short Introduction* (Oxford, 2002), 15.

13. Hansson, "Social Constructionism and Climate Science Denial," 4.

14. Alan Sokol and Jean Bricmont, "Defense of a Modest Scientific Realism," in *Knowledge and the World: Challenges Beyond the Science Wars*, ed. Martin Carrier, Johannes Roggenhofer, Günter Küppers, and Philippe Blanchard (Springer, 2004), 17.

15. Sokol and Bricmont, "Defense of a Modest Scientific Realism," 107.

16. Hansson, "Social Constructionism and Climate Science Denial," 5.

17. Hansson, "Social Constructionism and Climate Science Denial," 6.

18. Hansson, "Social Constructionism and Climate Science Denial," 7-8.

19. Mary Douglas and Aaron Wildavsky, *Risk and Culture: An Essay on the Selection of Technological and Environmental Dangers* (University of California Press, 1982); Hansson, "Social Constructionism and Climate Science Denial," 10.

20. Hansson, "Social Constructionism and Climate Science Denial," 17.

21. Robert C. Balling, *The Heated Debate: Greenhouse Predictions Versus Climate Reality* (Pacific Research Institute for Public Policy, 1992).

22. Hansson, "Social Constructionism and Climate Science Denial," 18.

23. Myanna Lahsen, "We Cannot Afford Not to Perform Constructionist Studies of Mainstream Climate Science," in *Climate Science and Society: A Primer*, ed. Zeke Baker, Tamar Law, Mark Vardy and Stephen Zehr (Earthscan Routledge, 2024), 29-38.

24. Harry Collins et al., "STS as Science or Politics?" *Social Studies of Science*, 47, no. 4 (2017): 580–586, 580, https://doi.org/10.1177/03063127177101

25. Lahsen, "We Cannot Afford Not to Perform Constructionist Studies," 31.

26. Bruno Latour, "Why Has Critique Run out of Steam? From Matters of Fact to Matters of Concern," *Critical Inquiry* 30, no. 2 (2004): 225–248. www.jstor.org/stable/10.1086/421123

27. Hans Radder, "Normative Reflexions on Constructivist Approaches to Science and Technology," *Social Studies of Science* 22, no. 1 (1992): 141–173, 156, http://www.jstor.org/stable/370230

28. Lahsen, "We Cannot Afford Not to Perform Constructionist Studies," 30.

29. Hansson, "Social Constructionism and Climate Science Denial."

30. Gregg Henrique, "Revisiting the Science Wars: Toward a Scientific Humanistic Worldview," *Psychology Today*, June 1, 2012, accessed November 4, 2024, https://www.psychologytoday.com/us/blog/theory-knowledge/201206/revisiting-the-science-wars

31. Jean Bricmont, "Jean Bricmont on Mara Beller," in *Knowledge and the World: Challenges Beyond the Science Wars*, ed. Martin Carrier, Johannes Roggenhofer, Günter Küppers, and Philippe Blanchard (Springer, 2004), 287.

32. Janny Scott, "Postmodern Gravity Deconstructed, Slyly," *New York Times*, May 18, 1996, https://www.nytimes.com/1996/05/18/nyregion/postmodern-gravity-deconstructed-slyly.html; Richard Rorty, "Phony Science Wars," *The Atlantic*, November 1999, https://www.theatlantic.com/magazine/archive/1999/11/phony-science-wars/377882/

33. Ava Kofman, "Bruno Latour, the Post-Truth Philosopher, Mounts a Defense of Science," *New York Times*, October 25, 2018, https://www.nytimes.com/2018/10/25/magazine/bruno-latour-post-truth-philosopher-science.html

34. Bricmont, "Jean Bricmont on Mara Beller," 286.

35. David Demeritt, "The Construction of Global Warming and the Politics of Science," *Annals of the Association of American Geographers* 91, no. 2 (2001): 307–337, 308–309, https://www.jstor.org/stable/3651262

36. Demeritt, "The Construction of Global Warming."

37. Quoted in Jop de Vrieze, "Science Wars' Veteran Has a New Mission," *Science*, October 10, 2017, https://www.science.org/content/article/bruno-latour-veteran-science-wars-has-new-mission

38. Valerie Richardson, "U.N. Mass Species Extinction Warning Blasted by Skeptics," *Washington Times*, May 22, 2019, https://www.washingtontimes.com/news/2019/may/22/un-mass-species-extinction-warning-blasted-skeptic/

39. Eva Lövbrand et al., "Who Speaks for the Future of Earth? How Critical Social Science Can Extend the Conversation on the Anthropocene," *Global Environmental Change* 32 (2015): 211–218, 211, https://doi.org/10.1016/j.gloenvcha.2015.03.012

40. David Victor, "Climate Change: Embed the Social Sciences in Climate Policy," *Nature* 520 (2015): 27–29, https://doi.org/10.1038/520027a

41. Victor, "Climate Change: Embed the Social Sciences."

42. Lövbrand et al., "Who Speaks for the Future," 214.

43. Lövbrand et al., "Who Speaks for the Future," 212.

44. Jeremy Baskin, "Paradigm Dressed as Epoch: The Ideology of the Anthropocene," *Environmental Values* 24, no. 1 (2015): 9-29, 10, 14, https://doi.org/10.3197/096327115X14183182353

45. Eduardo S. Brondizio et al., "Re-Conceptualizing the Anthropocene: A Call for Collaboration," *Global Environmental Change* 39 (2016): 318-327, 322, https://doi.org/10.1016/j.gloenvcha.2016.02.006

46. Lövbrand, et al., "Who Speaks for the Future," 212.

47. Baskin, "Paradigm Dressed as Epoch," 17.

48. "Wilderness Act," Pub. L. 88-577, 78 Stat. 890 (1964).

49. Baskin, "Paradigm Dressed as Epoch," 24.

50. Baskin, "Paradigm Dressed as Epoch," 24.

51. Nigel Clark and Yasmin Gunaratnam, "Earthing the *Anthropos*? From 'Socializing the Anthropocene' to Geologizing the Social," *European Journal of Social Theory* 20, no. 1 (2017): 146-163, 151, https://doi.org/10.1177/1368431016661337

52. Manuel Arias-Maldonado, "Bedrock or Social Construction? What Anthropocene Science Means for Political Theory," *Anthropocene Review* 7, no. 2 (2020): 97-112, 107, https://doi.org/10.1177/2053019619899

53. Gisli Palsson et al., "Reconceptualizing the 'Anthropos' in the Anthropocene: Integrating the Social Sciences and Humanities in Global Environmental Change Research," *Environmental Science and Policy* 28 (2013): 3-13, 7, https://doi.org/10.1016/j.envsci.2012.11.004

54. Will Steffen et al., "The Emergence and Evolution of Earth System Science," *Nature Reviews Earth and Environment* 1, no. 1 (2020): 54-63, https://doi.org/10.1038/s43017-019-0005-6

55. "Greenland Ice Cores Show Industrial Record of Acid Rain, Success of US Clean Air Act," *ScienceDaily*, accessed December 11, 2024, https://www.sciencedaily.com/releases/2014/04/140411091840.htm

56. Cade Metz et al., "Nobel Prize in Physics Awarded for Study of Humanity's Role in Changing Climate," *New York Times*, October 5, 2021, https://www.nytimes.com/2021/10/05/science/nobel-prize-physics-manabe-klaus-parisi.html

57. Noel Castree, "The 'Anthropocene' in Global Change Science: Expertise, the Earth, and the Future of Humanity," in *Anthropocene Encounters: New Directions in Green Political Thinking*, ed. Frank Biermann and Eva Lövbrand (Cambridge University Press, 2019), 25-49, 47.

58. Lövbrand et al., "Who Speaks for the Future," 216.

59. Latour, "Why Has Critique Run out of Steam."

60. Cited in Hansson, "Social Constructionism and Climate Science Denial," 2.

61. Manuel Arias-Maldonado and Zev Trachtenberg, eds. *Rethinking the Environment for the Anthropocene: Political Theory and Socionatural Relations in the New Geological Epoch* (Routledge, 2018), 6.

62. Radder, "Normative Reflexions on Constructivist Approaches."

Chapter 6

1. Jean Bricmont, "Jean Bricmont on Mara Beller," in *Knowledge and the World: Challenges Beyond the Science Wars,* ed. Martin Carrier, Johannes Roggenhofer, Günter Küppers, and Philippe Blanchard (Springer, 2004).

2. David Demeritt, "The Construction of Global Warming and the Politics of Science," *Annals of the Association of American Geographers* 91, no. 2 (2001): 307-337, 308-309, https://www.jstor.org/stable/3651262

3. Judith A. Layzer and Sara R. Rinfret, *The Environmental Case: Translating Values into Policy* (Sage/CQ Press, 2020), 6.

4. Anne Chung et al., "Translating Resilience-Based Management Theory to Practice for Coral Bleaching Recovery in Hawai'i," *Marine Policy* 99 (2019): 58-68, doi:10.1016/j.marpol.2018.10.013

5. Walter A. Rosenbaum, *Environmental Politics and Policy* (Sage/CQ Press, 2020), 60.

6. Michael D. Mastrandrea et al., "Guidance Note for Lead Authors of the IPCC Fifth Assessment Report on Consistent Treatment of Uncertainties," *Intergovernmental Panel on Climate Change (IPCC),* accessed November 4, 2024, https://www.ipcc.ch/site/assets/uploads/2018/03/inf09_p32_draft_Guidance_notes_LA_Consistent_Treatment_of_Uncertainties.pdf

7. Peter Jacobs et al., "The Arctic Is Now Warming Four Times as Fast as the Rest of the Globe," AGU Fall Meeting, December 2021, accessed November 4, 2024, https://agu.confex.com/agu/fm21/meetingapp.cgi/Paper/898204

8. Jacobs et al., "The Arctic Is Now Warming Four Times as Fast."

9. IPCC, "IPCC Special Report on the Ocean and Cryosphere in a Changing Climate: Summary for Policymakers," ed. H.-O. Pörtner et al., 2019, accessed November 4, 2024, https://www.ipcc.ch/srocc/chapter/summary-for-policymakers/

10. Keynyn Brysse et al., "Climate Change Prediction: Erring on the Side of Least Drama?," *Global Environmental Change* 23, no. 1 (2013): 327-337, https://doi.org/10.1016/j.gloenvcha.2012.10.008

11. Patrick Michaels and Sterling Burnett, "Climate Data Is Being Misused and Manipulated, Says Award-Winning Scientist," *Heartland Institute,* December 5, 2019, accessed November 4, 2024, https://www.heartland.org/news-opinion/news/it-takes-courage-to-stand-up-for-a-rational-discussion-of-climate-science-and-policy-says-award-winning-scientist

12. Will Wade, "Climate Impact 'Worse Than We Thought,' UN's Guterres Says," *Bloomberg Law,* September 16, 2021, accessed November 4, 2024, https://news.bloomberglaw.com/environment-and-energy/climate-impact-worse-than-we-thought-uns-guterres-says

13. Gernot Wagner and Martin L. Weitzman, *Climate Shock: The Economic Consequences of a Hotter Planet* (Princeton University Press, 2016).

14. Michael Mann, "The 'Fat Tail' of Climate Change Risk," *HuffPost,* September 11, 2015, https://www.huffpost.com/entry/the-fat-tail-of-climate-change-risk_b_8116264

15. Mark Cliffe, "The Sting of Climate Risk Is in the Tails," *Project Syndicate,*

September 16, 2021, accessed November 4, 2024, https://www.project-syndicate
.org/commentary/extreme-weather-fat-tail-risks-climate-change-by-mark-cliffe
-2022-09

16. Rosenbaum, *Environmental Politics*, 57.

17. Zachary A. Smith, *The Environmental Policy Paradox* (Routledge, 2018), 20.

18. "Rachel Carson Centennial," *John F. Kennedy Presidential Library and
Museum,* June 2, 2007, accessed November 4, 2024, https://www.jfklibrary.org/
events-and-awards/kennedy-library-forums/past-forums/transcripts/rachel-car
son-centennial#:~:text=I%20believe%20that%20Silent%20Spring,That's%20a%
20very%20big%20statement

19. "The Personal Attacks on Rachel Carson as a Woman Scientist," *Environment & Society Portal,* accessed November 4, 2024, https://www.environmentand
society.org/exhibitions/rachel-carsons-silent-spring/personal-attacks-rachel
-carson-woman-scientist

20. Cited in Vera Norwood, *Made from This Earth: American Women and Nature*
(University of North Carolina Press, 1993).

21. "The Desolate Year—Monsanto Magazine (1962)," *International Society for
Environmental Ethics (ISEE),* accessed November 4, 2024, https://enviroethics.org
/2011/12/02/the-desolate-year-monsanto-magazine-1962/

22. Anthony Leiserowitz et al., "Climate Change in the American Mind: Beliefs
& Attitudes, Spring 2023," *Yale Program on Climate Change Communication,* accessed November 4, 2024, https://climatecommunication.yale.edu/publications/
climate-change-in-the-american-mind-beliefs-attitudes-spring-2023/toc/2/

23. John Cook et al., "Quantifying the Consensus on Anthropogenic Global
Warming in the Scientific Literature," *Environmental Research Letters* 8, no. 2
(2013): 024024, https://doi.org/10.1088/1748-9326/8/2/024024

24. Cited in Oliver Burkeman, "Memo Exposes Bush's New Green Strategy,"
The Guardian, March 3, 2003, https://www.theguardian.com/environment/2003
/mar/04/usnews.climatechange

25. "Enact Ohio Higher Education Enhancement Act," Senate Bill 83, 135th
General Assembly, 2023-2024, https://legiscan.com/OH/bill/SB83/2023

26. "Intellectual Diversity in Higher Education Act Exposed," *ALEC Exposed:
Center for Media and Democracy,* accessed November 4, 2024, https://www.alecex
posed.org/wiki/Intellectual_Diversity_in_Higher_Education_Act_Exposed

27. "Enact Ohio Higher Education Enhancement Act," Ohio Legislature, 135th
General Assembly, accessed November 4, 2024, https://www.legislature.ohio.gov/
legislation/135/sb83

28. Alejandro De La Garza, "More States Want Students to Learn About Climate Science. Ohio Disagrees," *Time* (magazine), March 29, 2023, https://time.
com/6266938/ohio-climate-change-education-bill-culture-war/

29. "Ohio Senate Debates Bill That Would Force Universities to Give Weight to
Climate Denialism," *Columbia Climate School,* accessed November 4, 2024, https:/
/climate.law.columbia.edu/content/ohio-senate-debates-bill-would-force
-universities-give-weight-climate-denialism

30. "Regards State Higher Ed Institution Commitment to Certain Beliefs," Ohio Legislature,135th General Assembly, HB 394, accessed November 4, 2024, https://www.legislature.ohio.gov/legislation/135/hb394

31. Robert L. Heilbroner, *An Inquiry into the Human Prospect: With Second Thoughts and What Has Posterity Ever Done for Me?* (W. W. Norton & Company, 1975); William Ophuls, "Leviathan or Oblivion?," in *Toward a Steady-State Economy,* ed. Herman H. Daly (W. H. Freeman, 1973).

32. Dan Coby Shahar, "Rejecting Eco-Authoritarianism, Again," *Environmental Values* 24, no. 3 (2015): 345-366, https://www.jstor.org/stable/43695234

33. Mayson Glenn Obrien, "Authoritarian Environmentalism, Democracy, and Political Legitimacy," *Yale Review of International Studies,* March 10, 2020, accessed November 4, 2024, https://yris.yira.org/essays/authoritarian-environ mentalism-democracy-and-political-legitimacy/

34. Nico Stehr, "Exceptional Circumstances: Does Climate Change Trump Democracy?" *Issues in Science and Technology* 32, no. 2 (2016), accessed November 4, 2024, https://issues.org/exceptional-circumstances-does-climate-change-trump -democracy/

35. Stehr, "Exceptional Circumstances."

36. Stehr, "Exceptional Circumstances."

37. Daniel J. Fiorino, *Can Democracy Handle Climate Change?* (Polity, 2018).

38. Andrea Westall, "Exploring the Tensions: The Relationship Between Democracy and Sustainable Development," *Foundation for Democracy and Sustainable Development,* February 2, 2023, accessed November 4, 2024, https://www. fdsd.org/publications/the-relationship-between-democracy-and-sustainable -development/

39. Dan Banik, "Democracy and Sustainable Development," *Anthropocene Science* 1, no. 2 (2022): 233-245, https://doi.org/10.1007/s44177-022-00019-z

40. Frank Biermann, *Earth System Governance: World Politics in the Anthropocene* (MIT Press, 2014); John S. Dryzek, "Institutions for the Anthropocene: Governance in a Changing Earth System," *British Journal of Political Science* 46, no. 4 (2016): 937-956, doi:10.1017/S0007123414000453; Eva Lövbrand et al., "Earth System Governmentality: Reflections on Science in the Anthropocene," *Global Environmental Change* 19, no. 1 (2009): 7-13, https://doi.org/10.1016/j.gloenvcha.2008.10.002; Gerard Delanty and Aurea Mota, "Governing the Anthropocene: Agency, Governance, Knowledge," *European Journal of Social Theory* 20, no. 1 (2017): 9-38, https://doi.org/10.1177/ 1368431016668535; Amanda Machin, "Democracy and Agonism in the Anthropocene: The Challenges of Knowledge, Time and Boundary," *Environmental Values* 28, no. 3 (2019): 347-365, https://doi.org/10.3197/096327119X15519764179836

41. David Budtz Pedersen, "The Political Epistemology of Science-Based Policy-Making," *Society* 51, no. 5 (2014): 547-551, 547, https://doi.org/10.1007/s12115-014 -9820-z

42. Chris Ansell and Alison Gash, "Collaborative Governance in Theory and Practice," *Journal of Public Administration Research and Theory* 18, no. 4 (2007): 543-571, https://doi.org/10.1093/jopart/mum032

43. Niklas Wagner et al., "Effectiveness Factors and Impacts on Policymaking of Science-Policy Interfaces in the Environmental Sustainability Context," *Environmental Science and Policy* 140 (2023): 56-67, https://doi.org/10.1016/j.envsci.20 22.11.008

44. Wagner et al., "Effectiveness Factors."

45. Wagner et al., "Effectiveness Factors."

46. Daniel J. Fiorino, "Citizen Participation and Environmental Risk: A Survey of Institutional Mechanisms," *Science, Technology, and Human Values* 15, no. 2 (1990): 226-243, https://doi.org/10.1177/016224399001500204

47. Alice Dantas Brites et al., "Science-Based Stakeholder Dialogue for Environmental Policy Implementation," *Conservation & Society* 19, no. 4 (2021): 225-35, https://www.jstor.org/stable/27081510

48. Herman A. Karl et al., "A Dialogue, Not a Diatribe: Effective Integration of Science and Policy Through Joint Fact Finding," *Environment: Science and Policy for Sustainable Development* 49, no. 1 (2007): 20-34, https://doi.org/10.3200/ENVT.49.1 .20-34

49. Rosenbaum, *Environmental Politics.*

50. Lawrence E. Susskind et al., "Arguing, Bargaining and Getting Agreement," in *The Oxford Handbook of Public Policy*, ed. Michael Moran, Martin Rein, and Robert E. Goodin (Oxford University Press, 2005), 269-295.

Chapter 7

1. "History of Energy Consumption in the United States, 1775-2009," *Energy Information Administration*, February 9, 2011, accessed November 4, 2024, https:/ /www.eia.gov/todayinenergy/detail.php?id=10

2. "Guterres Calls for Phasing out Fossil Fuels to Avoid Climate 'Catastrophe,'" *UN News*, June 15, 2023, accessed November 4, 2024, https://news.un.org/en/ story/2023/06/1137747

3. Paul J. Crutzen, "Geology of Mankind," *Nature* 415, no. 23 (2002), https://doi .org/10.1038/415023a

4. Mark G. Lawrence and Paul J. Crutzen, "Was Breaking the Taboo on Research on Climate Engineering via Albedo Modification a Moral Hazard, or a Moral Imperative?" *Earth's Future* 5, no. 2 (2017): 136-143, 138, https://doi.org/10 .1002/2016EF000463

5. Paul J. Crutzen, "Albedo Enhancement by Stratospheric Sulfur Injections: A Contribution to Resolve a Policy Dilemma?" *Climatic Change* 77 (2006): 211-220, 211, https://doi.org/10.1007/s10584-006-9101-y

6. Christian Schwaegerl, "The Anthropocene: Paul Crutzen's Epochal Legacy," *Anthropocene Magazine*, February 14, 2021, accessed November 4, 2024, https:// www.anthropocenemagazine.org/2021/02/the-anthropocene-paul-crutzens -epochal-legacy/

7. P. R. Shukla et al., eds. "IPCC, 2022: Summary for Policymakers," in *Climate Change 2022: Mitigation of Climate Change*, Contribution of Working Group III to

the Sixth Assessment Report of the Intergovernmental Panel on Climate Change (Cambridge University Press, 2022), doi: 10.1017/9781009157926.001

8. James Temple, "What Is Geoengineering—and Why Should You Care?" *MIT Technology Review*, August 9, 2019, accessed November 4, 2024, https://www.technologyreview.com/2019/08/09/615/what-is-geoengineering-and-why-should-you-care-climate-change-harvard/

9. "New Report Says U.S. Should Cautiously Pursue Solar Geoengineering Research to Better Understand Options for Responding to Climate Change Risks," *National Academies*, March 25, 2021, accessed November 4, 2024, https://www.nationalacademies.org/news/2021/03/new-report-says-u-s-should-cautiously-pursue-solar-geoengineering-research-to-better-understand-options-for-responding-to-climate-change-risks

10. James E. Hansen et al., "Global Warming in the Pipeline," *Oxford Open Climate Change* 3, no. 1 (2023): kgad008, https://doi.org/10.1093/oxfclm/kgad008

11. Holly Buck, *After Geoengineering: Climate Tragedy, Repair, and Restoration* (Verso Books, 2019).

12. A. Atiq Rahman et al., "Developing Countries Must Lead on Solar Geoengineering Research," *Nature* 556, no. 7699 (2018): 22–24, https://doi.org/10.1038/d41586-018-03917-8

13. Michael Mann and Tom Toles, *The Madhouse Effect: How Climate Change Denial Is Threatening Our Planet, Destroying Our Politics, and Driving Us Crazy* (Columbia University Press, 2016), 128.

14. H.-O. Pörtner et al., "Summary for Policymakers," in *Climate Change 2022: Impacts, Adaptation and Vulnerability,* Contribution of Working Group II to the Sixth Assessment Report of the Intergovernmental Panel on Climate Change (Cambridge University Press, 2022), 3–33, doi:10.1017/9781009325844.001

15. Kate Marvel, "A Handful of Dust," in *All We Can Save: Truth, Courage, and Solutions for the Climate Crisis,* ed. Ayana Elizabeth Johnson and Katharine K. Wilkinson (One World, 2020), 30–35, 35.

16. Elizabeth Kolbert, *Under a White Sky: The Nature of the Future* (Crown, 2022).

17. IPCC, *Climate Change 2021: The Physical Science Basis*, Contribution of Working Group I to the Sixth Assessment Report of the Intergovernmental Panel on Climate Change, ed. Valérie Masson-Delmotte et al. (Cambridge University Press, 2023), https://doi.org/10.1017/9781009157896

18. Astrid Dannenberg and Sonja Zitzelsberger, "Climate Experts' Views on Geoengineering Depend on Their Beliefs About Climate Change Impacts," *Nature Climate Change* 9 (2019): 769–775, https://doi.org/10.1038/s41558-019-0564-z

19. Astrid Dannenberg et al., "Climate Negotiators' and Scientists' Assessments of the Climate Negotiations," *Nature Climate Change* 7, no. 6 (2017): 437–442, https://doi.org/10.1038/nclimate3288; Dannenberg et al. report the majority of the respondents in their sample majored in "natural sciences (37 percent), followed by economics and business administration (17 percent), and engineering (14 percent)."

20. Sara Stefanini, "Switzerland Puts Geoengineering Governance on UN Environment Agenda," *Climate Home News*, February 26, 2019, accessed November 4, 2024, https://www.climatechangenews.com/2019/02/26/swiss-push-talk-geoengineering-goes-sci-fi-reality/

21. Pörtner et al., *Climate Change 2022: Impacts, Adaptation, and Vulnerability.*

22. Sikina Jinnah and Simon Nicholson, "The Hidden Politics of Climate Engineering," *Nature Geoscience* 12, no. 11 (2019): 876–879, https://doi.org/10.1038/s41561-019-0483-7

23. Christopher Flavelle, "The U.S. Is Building an Early Warning System to Detect Geoengineering," *New York Times*, November 28, 2024, https://www.nytimes.com/2024/11/28/climate/geoengineering-early-warning-system.html

24. Janos Pasztor, "Calls for an SRM 'Non-Use Agreement' Underline the Need for Governance," *C2G*, January 28, 2022, accessed November 4, 2024, https://www.c2g2.net/calls-for-an-srm-non-use-agreement/; see also: "Solar Geoengineering Non-Use Agreement," accessed November 4, 2024, https://www.solargeoeng.org/

25. Jennie C. Stephens and Kevin Surprise, "The Hidden Injustices of Advancing Solar Geoengineering Research," *Global Sustainability* 3 (2020): 3:e2, https://doi.org/10.1017/sus.2019.28

26. Kevin Anderson et al., "Controversies of Carbon Dioxide Removal," *Nature Reviews Earth and Environment* 4 (2023): 808–814, https://doi.org/10.1038/s43017-023-00493-y

27. Anderson et al., "Controversies."

28. Shukla et al., *Climate Change 2022: Mitigation of Climate Change.*

29. "Carbon Capture and Storage: Actions Needed to Improve DOE Management of Demonstration Projects," *Government Accountability Office*, GAO-22-105111, December 20, 2021, https://www.gao.gov/products/gao-22-105111

30. "Inflation Reduction Act," Pub. L. 117-169, 136 Stat. 1818 (2022).

31. Kristoffer Tigue, "How Midwest Landowners Helped to Derail One of the Biggest CO_2 Pipelines Ever Proposed," *Inside Climate News*, November 5, 2023, accessed November 4, 2024, https://insideclimatenews.org/news/05112023/landowners-fight-co2-pipeline-midwest-navigator/

32. "Ban on Carbon Capture & Sequestration (CCS) Facilities & Pipelines Passed by Full New Orleans City Council," *Alliance for Affordable Energy*, June 9, 2022, accessed November 4, 2024, https://www.all4energy.org/watchdog/ban-on-carbon-capture-sequestration-ccs-facilities-pipelines-passed-by-full-new-orleans-city-council/

33. Tigue, "How Midwest Landowners."

34. Tigue, "How Midwest Landowners."

35. Tigue, "How Midwest Landowners."

36. David Roberts, "House Democrats Just Put out the Most Detailed Climate Plan in US Political History," *Vox*, June 30, 2020, accessed November 4, 2024, https://www.vox.com/energy-and-environment/2020/6/30/21305891/aoc-climate-change-house-democrats-select-committee-report

37. Brian Kahn, "Republicans Decided to Talk About a Dramatic Climate Solution Without Addressing the Actual Problem," *Gizmodo,* November 8, 2017, accessed November 4, 2024, https://gizmodo.com/republicans-decided-to-talk-about-a-dramatic-climate-so-1820266491

38. Kahn, "Republicans Decided."

39. Nick Sobczyk, "Red New Deal? GOP Offers a Climate Plan of Its Own," *E&E News*, November 4, 2021, accessed November 4, 2024, https://www.eenews.net/articles/red-new-deal-gop-offers-a-climate-plan-of-its-own/

40. Mann and Toles, *The Madhouse Effect,* 118.

41. Anne Fremaux, "The Return of Nature in the Capitalocene: A Critique of the Ecomodernist Version of the 'Good Anthropocene,'" in *Rethinking the Environment for the Anthropocene: Political Theory and Socionatural Relations in the New Geological Epoch*, ed. Manuel Arias-Maldonado and Zev Trachtenberg (Routledge, 2019), 19–36, 21.

42. "Fuel to the Fire: How Geoengineering Threatens to Entrench Fossil Fuels and Accelerate the Climate Crisis," *Center for International Environmental Law* (2019): 1–71, 42–43, accessed November 4, 2024, https://www.ciel.org/wp-content/uploads/2019/02/CIEL_FUEL-TO-THE-FIRE_How-Geoengineering-Threatens-to-Entrench-Fossil-Fuels-and-Accelerate-the-Climate-Crisis_February-2019.pdf

43. Jean-Daniel Collomb, "US Conservative and Libertarian Experts and Solar Geoengineering: An Assessment," *European Journal of American Studies* 14, no. 2 (2019), https://doi.org/10.4000/ejas.14717

44. *Massachusetts v. EPA,* 549 U.S. 497 (2007).

45. The eleven states joining Massachusetts in the lawsuit included California, Connecticut, Illinois, Maine, New Jersey, New Mexico, New York, Oregon, Rhode Island, Vermont, and Washington. Accessed November 4, 2024, https://ballotpedia.org/Massachusetts_v._Environmental_Protection_Agency

46. *Massachusetts v. EPA,* 549 U.S. 497 (2007).

47. Lee Lane, "Strategic Options for Bush Administration Climate Policy" (AEI Press, 2006), accessed November 4, 2024, https://www.aei.org/wp-content/uploads/2014/07/-strategic-options-for-the-bush-administration_160635893399.pdf

48. Lane, "Strategic Options for Bush Administration Climate Policy," 70.

49. "Geo-Engineering Seen as a Practical, Cost-Effective Global Warming Strategy," *Heartland Institute*, December 1, 2007, accessed November 4, 2024, https://heartland.org/opinion/geo-engineering-seen-as-a-practical-cost-effective-global-warming-strategy/

50. John M. Broder, "Obama Affirms Climate Change Goals," *New York Times*, November 18, 2009, https://www.nytimes.com/2008/11/19/us/politics/19climate.html

51. "Geoengineering: A Revolutionary Approach to Climate Change," *American Enterprise Institute*, accessed November 4, 2024, https://www.aei.org/events/geoengineering-a-revolutionary-approach-to-climate-change/

52. Lee Lane, "Researching Solar Geoengineering as a Climate Policy Option,"

House Committee on Science and Technology, November 5, 2009, accessed November 4, 2024, https://www.researchgate.net/publication/325877288_State ment_of_Lee_Lane_resident_fellow_co-director_AEI_geoengineering_project_ American_enterprise_institute_before_the_house_committee_on_science_and _technology_hearing_on_researching_solar_radiation_m

53. "American Clean Energy and Security Act," H.R. 2454, 111th Cong. (2009-2010).

54. Environmental Protection Agency, "Endangerment and Cause or Contribute Findings for Greenhouse Gases Under Section 202(a) of the Clean Air Act," *Federal Register* 74, no. 239 (December 15, 2009): 66496, https://www.govinfo.gov /content/pkg/FR-2009-12-15/pdf/E9-29537.pdf

55. Diana Furchtgott-Roth, "Climate Change: Another Option," *Manhattan Institute,* December 10, 2009, accessed November 21, 2024, https://manhattan. institute/article/climate-change-another-option

56. Furchtgott-Roth, "Climate Change: Another Option."

57. Collomb, "US Conservative and Libertarian Experts."

58. Environmental Protection Agency, "Carbon Pollution Emission Guidelines for Existing Stationary Sources: Electric Utility Generating Units," *Federal Register* 80, no. 205 (October 23, 2015): 64662, https://www.govinfo.gov/content/pkg/ FR-2015-10-23/pdf/2015-22842.pdf

59. David R. Henderson, "Egad, Engineering!" *Cato Institute* (Winter 2015-2016), accessed November 4, 2024, https://www.cato.org/regulation/winter-2015 -2016/climate-shock#

60. Benjamin Hulac and Jean Chemnick, "For Tillerson, Climate Change Is an 'Engineering Problem,'" *E&E News by Politico,* December 12, 2016, accessed November 4, 2024, https://www.eenews.net/articles/for-tillerson-climate-change-is an-engineering-problem/

61. Collomb, "US Conservative and Libertarian Experts."

62. "Fuel to the Fire," 45.

63. "Fuel to the Fire," 45.

64. Benjamin Franta and Geoffrey Supran, "The Fossil Fuel Industry's Invisible Colonization of Academia," *The Guardian,* March 13, 2017, https://www.theguar dian.com/environment/climate-consensus-97-per-cent/2017/mar/13/the-fossil -fuel-industrys-invisible-colonization-of-academia

65. "Funding," *Harvard's Solar Geoengineering Research Program,* accessed November 4, 2024, https://geoengineering.environment.harvard.edu/funding

66. Zack Budryk, "More Than 500 Academics Call on Universities to Stop Accepting Research Funding from Fossil Fuel Industry," *The Hill,* March 22, 2022, accessed November 4, 2024, https://thehill.com/policy/energy-environment/ 599206-more-than-500-academics-call-on-universities-to-stop-accepting/

67. Fiona Harvey, "Universities Must Reject Fossil Fuel Cash for Climate Research, Say Academics," *The Guardian,* March 21, 2022, https://www.theguardian .com/science/2022/mar/21/universities-must-reject-fossil-fuel-cash-for-climate -research-say-academics

68. Katherine Richardson et al., "Earth Beyond Six of Nine Planetary Boundaries," *Science Advances* 9, no. 37 (2023), doi:10.1126/sciadv.adh2458

69. Sean Mowbray, "Beyond Climate: Oil, Gas and Coal Are Destabilizing All 9 Planetary Boundaries," *Mongabay,* November 14, 2023, accessed November 4, 2024, https://news.mongabay.com/2023/11/beyond-climate-oil-gas-and-coal-are -destabilizing-all-9-planetary-boundaries/

70. "United States Produces More Crude Oil Than Any Country, Ever," *Energy Information Administration,* March 11, 2024, accessed January 3, 2025, https:// www.eia.gov/todayinenergy/detail.php?id=61545

71. Rachel Maddow, *Blowout: Corrupted Democracy, Rogue State Russia, and the Richest, Most Destructive Industry on Earth* (Crown, 2019).

72. Bill McKibben, "Global Warming's Terrifying New Math," *Rolling Stone,* July 19, 2012, https://www.rollingstone.com/politics/politics-news/global-warm ings-terrifying-new-math-188550/

73. Bill McKibben, "Recalculating the Climate Math," *New Republic,* September 22, 2016, https://newrepublic.com/article/136987/recalculating-climate-math

74. "Mitigation of Climate Change Report 2022: "Litany of Broken Climate Promises," *United Nations,* 5 min. 12 sec., accessed November 4, 2024, https:// www.youtube.com/watch?v=P8rlLaT8v4Q

75. "Inflation Reduction Act."

76. Regional Greenhouse Gas Initiative, https://www.rggi.org/

77. "Investment of Proceeds," *Regional Greenhouse Gas Initiative,* accessed November 4, 2024, https://www.rggi.org/investments/proceeds-investments

78. William J. Ripple et al., "The 2023 State of the Climate Report: Entering Uncharted Territory," *BioScience* 73, no. 12 (2023): 841-850, https://doi.org/10 .1093/biosci/biad080

79. "The Evidence Is Clear: The Time for Action Is Now. We Can Halve Emissions by 2030," *IPCC Newsroom,* April 4, 2022, accessed November 4, 2024, https: //www.ipcc.ch/2022/04/04/ipcc-ar6-wgiii-pressrelease/

Chapter 8

1. Corey J. A. Bradshaw et al., "Underestimating the Challenges of Avoiding a Ghastly Future," *Frontiers in Conservation Science* 1 (2021), https://doi.org/10.3389 /fcosc.2020.615419

2. Aldo Leopold, *A Sand County Almanac, with Other Essays on Conservation from Round River* (Oxford University Press, 1966).

3. E. F. Schumacher, *A Guide for the Perplexed* (Jonathan Cape Ltd., 1977), 152-153.

4. David Ehrenfeld, *The Arrogance of Humanism* (Oxford University Press, 1978), 236-237.

5 Robin Wall Kimmerer, *Braiding Sweetgrass: Indigenous Wisdom, Scientific Knowledge and the Teachings of Plants* (Milkweed Editions, 2013).

6. Jeffrey M. Jones, "Four in 10 Americans Say They Are Environmentalists," *Gallup,* April 21, 2021, accessed October 31, 2024, https://news.gallup.com/poll/

348227/one-four-americans-say-environmentalists.aspx#:~:text=When%20Gal lup%20first%20asked%20this,three%20polls%20conducted%20since%202016

7. Anthony Leiserowitz et al., "Dramatic Increase in Public Beliefs and Worries About Climate Change," *Yale Program on Climate Change Communication*, September 7, 2021, accessed October 31, 2024, https://climatecommunication.yale.edu/publications/dramatic-increase-in-public-beliefs-and-worries-about-climate-change/

8. Mike Hulme, *Why We Disagree About Climate Change: Understanding Controversy, Inaction and Opportunity* (Cambridge University Press, 2009), 189-190.

9. Andrew J. Hoffman, *How Culture Shapes the Climate Change Belief* (Stanford University Press, 2015), 3-4.

10. Rebecca Solnit Quotes, accessed November 6, 2024, https://quotefancy.com /rebecca-solnit-quotes

11. Anthony Leiserowitz et al., "Climate Change in the American Mind: Beliefs & Attitudes, Spring 2023," *Yale Program on Climate Change Communication*, accessed November 4, 2024, https://climatecommunication.yale.edu/publications/climate-change-in-the-american-mind-beliefs-attitudes-spring-2023/toc/2/

12. Cited in Oliver Burkeman, "Memo Exposes Bush's New Green Strategy," *The Guardian*, March 3, 2003, https://www.theguardian.com/environment/2003/mar/04/usnews.climatechange

13. Shaun W. Elsasser and Riley E. Dunlap, "Leading Voices in the Denier Choir: Conservative Columnists' Dismissal of Global Warming and Denigration of Climate Science," *American Behavioral Scientist* 57, no. 6 (2013): 754-776, https://doi .org/10.1177/0002764212469800

14. Sander van der Linden, "The Gateway Belief Model (GBM): A Review and Research Agenda for Communicating the Scientific Consensus on Climate Change," *Current Opinion in Psychology* 42 (2021): 7-12, 7, https://doi.org/10.1016/j.copsyc.2021.01.005

15. Matthew H. Goldberg et al., "The Role of Anchoring in Judgments About Expert Consensus," *Journal of Applied Social Psychology* 49, no. 3 (2019): 192-200, 192, https://doi.org/10.1111/jasp.12576

16. Goldberg, "The Role of Anchoring," 192.

17. Sander van der Linden et al., "The Gateway Belief Model: A Large-Scale Replication," *Journal of Environmental Psychology* 62 (2019): 49-58, https://doi.org /10.1016/j.jenvp.2019.01.009

18. Stephan Lewandowsky et al., "The Pivotal Role of Perceived Scientific Consensus in Acceptance of Science," *Nature Climate Change* 3, no. 4 (2013): 399-404, https://doi.org/10.1038/nclimate1720

19. Cited in Burkeman, "Memo Exposes Bush's New Green Strategy."

20. Leiserowitz et al., "Climate Change in the American Mind."

21. Gregg Sparkman et al., "Americans Experience a False Social Reality by Underestimating Popular Climate Policy Support by Nearly Half," *Nature Communications* 13, no. 4779 (2022), https://doi.org/10.1038/s41467-022-32412-y

22. Nathaniel Geiger and Janet K. Swim, "Climate of Silence: Pluralistic Igno-

rance as a Barrier to Climate Change Discussion," *Journal of Environmental Psychology* 47 (2016): 79–90, https://doi.org/10.1016/j.jenvp.2016.05.002

23. Edward Maibach et al., "Is There a Climate "Spiral of Silence in America?" *Yale Program on Climate Change Communication,* September 29, 2016, accessed November 6, 2024, https://climatecommunication.yale.edu/publications/climate-spiral-silence-america/

24. Matthew H. Goldberg et al., "Perceived Social Consensus Can Reduce Ideological Biases on Climate Change," *Environment and Behavior* 52, no. 5 (2019): 495–517, https://doi.org/10.1177/0013916519853302

25. Goldberg et al., "Perceived Social Consensus," 5.

26. Francis Commerçon et al., "Radio Stories Increase Conservatives' Beliefs That Republicans Are Worried About Climate Change," *Yale Program on Climate Change Communication,* July 21, 2021, accessed November 6, 2024, https://climatecommunication.yale.edu/publications/radio-stories-increase-conservatives-beliefs-that-republicans-are-worried-about-climate-change/

27. republicEn, https://republicen.org/

28. republicEn.

29. Abel Gustafson et al., "Personal Stories Can Shift Climate Change Beliefs and Risk Perceptions: The Mediating Role of Emotion," *Communication Reports* 33, no. 3 (2020): 121–135, https://doi.org/10.1080/08934215.2020.1799049

30. Gustafson et al., "Personal Stories," 15.

31. Annie Sneed, "Conservative Hunters and Fishers May Help Determine the Fate of National Monuments," *Scientific American,* October 30, 2017, accessed November 6, 2024, https://www.scientificamerican.com/article/conservative-hunters-and-fishers-may-help-determine-the-fate-of-national-monuments1/

32. "On the Front Line of Wildlife Decline," *NWF Outdoors Team,* March 30, 2022, accessed January 3, 2025, https://www.nwf.org/Outdoors/Blog/03-30-2022-Habitat-Loss#:~:text=During%20the%201800s%2C%20the%20U.S.,supporting%20ethical%2C%20regulated%20hunting%20practices

33. Livia Gershon, "Rachel Carson's Critics Called Her a Witch," *JSTOR Daily,* February 21, 2019, accessed November 6, 2024, https://daily.jstor.org/rachel-carsons-critics-called-her-a-witch/

34. Gershon, "Rachel Carson's Critics."

35. "*Silent Spring* in Popular Culture," *Environment & Society Portal,* accessed November 6, 2024, https://www.environmentandsociety.org/exhibitions/rachel-carsons-silent-spring/silent-spring-popular-culture

36. Roy Robertson, "Where Have All the Quail Gone?" *Farm Progress,* March 8, 2008, accessed November 6, 2024, https://www.farmprogress.com/commentary/where-have-all-the-quail-gone-

37. "Conservation Hawks," accessed November 6, 2024, https://www.conservationhawks.org

38. "Secretary Zinke, Stand Up for Public Lands," *YouTube,* 0:30, accessed November 6, 2024, https://www.youtube.com/watch?v=eMay5BVBPEE

39. Sneed, "Conservative Hunters and Fishers."

40. Sneed, "Conservative Hunters and Fishers."

41. Roger Fisher and William L. Ury, *Getting to Yes: Negotiating Agreement Without Giving In* (Penguin Books, 1983).

42. My original advisor during my doctoral studies mediated this dispute and shared the story with students.

43. "The EcoRight," *republicEn*, accessed November 6, 2024, https://republicen .org/ecoright

44. "Our Endangered Values, by Jimmy Carter: on Environment," accessed November 6, 2024, https://ontheissues.org/Archive/Endangered_Values_Environ ment.htm

45. "Alaska National Interest Lands Conservation Act," Pub. L. No. 96–487, 94 Stat. 2371.

46. David W. Moore, "Public Opposes Oil Drilling in ANWR," *Gallup*, April 24, 2002, accessed November 6, 2024, https://news.gallup.com/poll/5884/public -opposes-oil-drilling-anwr.aspx

47. "Tax Cuts and Jobs Act of 2017," Pub. L. No. 115–97, 131 Stat. 2054.

48. Matthew Ballew et al., "Americans Oppose Drilling in the ANWR," *Yale Program on Climate Change Communication*, September 26, 2019, accessed November 6, 2024, https://climatecommunication.yale.edu/publications/americans-op pose-drilling-arctic-national-wildlife-refuge-2019/

49. Sophie Lewis, "Biden Administration Suspends Trump-Era Oil and Gas Drilling Leases in Arctic National Wildlife Refuge," *CBS News*, June 2, 2021, accessed November 6, 2024, https://www.cbsnews.com/news/biden-suspends -trump-oil-gas-drilling-lease-arctic-national-wildlife-refuge-alaska/

50. "2023 Camping Report: Camping Demand Is at an All-Time High," *the dyrt*, accessed November 6, 2024, https://reports.thedyrt.com/2023-camping-report/

51. Adam Skolnick, "Why Environmentalists Are Reclaiming Patriotism," *Outside*, October 30, 2020, accessed November 6, 2024, https://www.outsideonline .com/2418261/environmentalists-reclaiming-patriotism-flag

52. Alison Chase, "Strong Support Exists for a Bold 30x30 Vision," *National Resources Defense Council* (expert blog), February 2, 2021, accessed November 6, 2024, https://www.nrdc.org/bio/alison-chase/strong-support-exists-bold-30x30 -vision

53. "Public Opinion Research Regarding 30x30," *ALG Research*, January 26, 2021, accessed November 6, 2024, https://www.nrdc.org/sites/default/files/nrdc _30x30_public_release_memo_.pdf

54. "National Park Wildlife Poll," *National Parks Conservation Association*, November 8, 2023, accessed November 6, 2024, https://www.npca.org/resources/ 3462-national-park-wildlife-poll

55. "National Park Wildlife Poll."

56. "National Park Wildlife Poll."

57. Ramsey Touchberry, "Plastic Is Everywhere Except in Democrats' Plans for a Fossil Fuel-Free Future," *Washington Times*, April 8, 2022, https://www.

washingtontimes.com/news/2022/apr/8/plastics-oil-product-omitted-dems-cli
mate-plans/

58. Tom Rogan, "Conservatives Should Support Action to Reduce Plastic Pollution," *Washington Examiner*, April 9, 2018, https://www.washingtonexaminer.com
/opinion/2590369/conservatives-should-support-action-to-reduce-plastic-pollu
tion/

59. Emily Becker and Mary Sagatelova, "TOPLINES: Polling on Clean Energy
in the 2024 Presidential Election," *Third Way*, February 14, 2024, accessed November 6, 2024, https://www.thirdway.org/polling/toplines-polling-on-clean-en
ergy-in-the-2024-presidential-election

60. Cary Funk and Meg Hefferon, "U.S. Public Views on Climate and Energy,"
Pew Research Center, November 25, 2019, accessed November 6, 2024, https://
www.pewresearch.org/science/2019/11/25/u-s-public-views-on-climate-and
-energy/

61. "Inflation Reduction Act," Pub. L. 117-169, 136 Stat. 1818 (2022).

62. Barry Rabe, "A New Era in States' Climate Policies?" in *Changing Climate
Politics: US Policies and Civic Action*, ed. Yael Wolinsky-Nahmias (CQ Press, 2014),
56.

63. "Texas: Profile Analysis," *Energy Information Administration*, July 18, 2024,
https://www.eia.gov/state/analysis.php?sid=TX

64. "Fact Checker: Does Gov. Kim Reynolds Deny Climate Change?" *The Ga-
zette*, March, 20, 2022, https://www.thegazette.com/government-politics/fact
-checker-does-gov-kim-reynolds-deny-climate-change/

65. "Renewable Energy Works for Iowa," *Iowa Environmental Council*, accessed
November 6, 2024, https://www.iaenvironment.org/our-work/clean-energy/
renewable-iowa

66. "American Clean Power Honors Iowa Champions of Clean Energy," *American Clean Power*, accessed November 6, 2024, https://cleanpower.org/news/acp
-honors-ia-champions-of-clean-energy/

67. Mitch Perry, "DeSantis Signs Bill Erasing the Term 'Climate Change' from
State Law," *Florida Phoenix*, May 15, 2024, https://floridaphoenix.com/2024/05/
15/desantis-signs-bill-erasing-the-term-climate-change-from-state-law/; Eric
Wesoff, "Florida Is Now Adding More Solar Power Than Any Other State," *Canary
Media*, September 15, 2023, accessed November 6, 2024, https://www.canarymedia
.com/articles/solar/florida-is-now-adding-more-solar-power-than-any-other-state

68. Barry G. Rabe, *Statehouse and Greenhouse: The Emerging Politics of American
Climate Change Policy* (Brookings Institution Press, 2004).

69. Adam Wren and Debra Kahn, "A Red-State Governor Walks into a COP,"
Politico, November 15, 2022, accessed November 7, 2024, https://www.politico.
com/newsletters/the-long-game/2022/11/15/a-red-state-governor-walks-into-a
-cop-00066950

70. "About Us," *Conservative Energy Network*, accessed November 6, 2024, https:
//conservativeenergynetwork.org/about-cen/

71. Yale Climate Connections Team, "Why a Conservative Nonprofit Is Promoting Renewable Energy," *Yale Climate Connections*, June 22, 2023, accessed November 6, 2024, http://yaleclimateconnections.org/2023/06/why-a-conservative-nonprofit-is-promoting-renewable-energy/

72. "Timeline of 3M History," *3M*, accessed November 6, 2024, https://www.3m.com/3M/en_US/company-us/about-3m/history/timeline/#:~:text=1975%20%E2%80%93%20Eliminating%20Pollution,saved%203M%20billions%20in%20costs; "Our Global Impact—Sustainability," *3M*, accessed November 6, 2024, https://www.3m.com/3M/en_US/sustainability-us/strategy/

73. Lucy Buchholz, "Top 10: ESG Strategies from the World's Largest Companies," *Sustainability Magazine*, July 7, 2023, accessed November 6, 2024, https://sustainabilitymag.com/top10/top-10

74. Christina DeConcini et al., "One Year In, How the Inflation Reduction Act Is Creating a Manufacturing Resurgence in the US," *World Resources Institute*, August 9, 2023, accessed November 6, 2024, https://www.wri.org/insights/inflation-reduction-act-anniversary-manufacturing-resurgence

75. "Food Waste FAQS," *U.S. Department of Agriculture*, accessed November 6, 2024, https://www.usda.gov/foodwaste/faqs

76. "Reduced Food Waste," *Project Drawdown*, accessed November 6, 2024, https://drawdown.org/solutions/reduced-food-waste

77. Drew DeSilver, "Americans Say They're Changing Behaviors to Help the Environment—but Is It Making a Difference?" *Pew Research Center* (blog), December 19, 2019, accessed November 6, 2024, https://www.pewresearch.org/short-reads/2019/12/19/americans-say-theyre-changing-behaviors-to-help-the-environment-but-is-it-making-a-difference/

78. Blaine Fulmer, "Environmentalism Isn't Partisan—At Least It Shouldn't Be," *State of the Planet, Columbia Climate School*, August 2, 2022, accessed November 6, 2024, https://news.climate.columbia.edu/2022/08/02/environmentalism-isnt-partisan-at-least-it-shouldnt-be/

79. Katharine Hayhoe, *Saving Us: A Climate Scientist's Case for Hope and Healing in a Divided World* (One Signal Publishers, 2021).

80. "Seven Things You Need to Know About Climate Change and Conflict," *International Committee of the Red Cross*, September 7, 2020, accessed November 6, 2024, https://www.icrc.org/en/document/climate-change-and-conflict#:~:text=Climate%20change%20does%20not%20directly,social%2C%20economic%20and%20environmental%20factors

81. "Five Ways the Climate Crisis Impacts Human Security," *United Nations*, accessed November 6, 2024, https://www.un.org/en/climatechange/science/climate-issues/human-security; see also John D. Banusiewicz, "Hagel to Address 'Threat Multiplier' of Climate Change," *U.S. Department of Defense*, October 13, 2014, accessed November 6, 2024, https://www.defense.gov/News/News-Stories/Article/Article/603440/hagel-to-address-threat-multiplier-of-climate-change/

82. "U.S. Billion-Dollar Weather and Climate Disasters," *NOAA National Cen-*

ters for Environmental Information (NCEI), accessed November 6, 2024, https://www.ncei.noaa.gov/access/billions/

83. Christopher Flavelle, "Climate Change Is Bankrupting America's Small Towns," *New York Times*, September 15, 2021, https://www.nytimes.com/2021/09/02/climate/climate-towns-bankruptcy.html

84. Alex Brown, "States Beg Insurers Not to Drop Climate-Threatened Homes," *Stateline*, June 5, 2024, accessed November 6, 2024, https://stateline.org/2024/06/05/states-beg-insurers-not-to-drop-climate-threatened-homes/

85. Kennedy Mason, "Florida Residents Being Dropped by Private Insurance Companies Turn to State-Backed Insurer," *WUFT, PBS Newshour*, March 7, 2023, accessed November 6, 2024, https://www.wuft.org/state-news/2023-03-07/flori da-residents-being-dropped-by-private-insurance-companies-turn-to-state -backed-insurer

86. Associated Press, "Florida, Other States Beg Insurers Not to Drop Climate-Threatened Homes," *Tampa Bay Times*, June 9, 2022, https://www.tampabay.com /news/business/2024/06/09/florida-other-states-beg-insurers-not-drop-climate -threatened-homes/

87. Kate Wheeling, "Climate Migration Has Come to the United States," *The Nation*, April 16, 2021, https://www.thenation.com/article/environment/califor nias-climate-migrants-fire/; Abrahm Lustgarten, "Climate Change Will Force a New American Migration," *ProPublica*, September 15, 2020, https://www.propub lica.org/article/climate-change-will-force-a-new-american-migration

88. Samantha Allen, "30% of Americans Cite Climate Change as a Motivator to Move in 2024," *Forbes*, January 5, 2024, https://www.forbes.com/home-improve ment/features/americans-moving-climate-change/

89. Aimee Picchi, "About 3 Million Americans Are Already 'Climate Migrants,' Analysis Finds. Here's Where They Left," *CBS News*, December 18, 2023, accessed November 6, 2024, https://www.cbsnews.com/news/climate-change-america-3 -million-migrants-first-street-nature/

90. Abrahm Lustgarten, *On the Move: The Overheating Earth and the Uprooting of America* (Farrar, Straus and Giroux, 2024).

91. Qi Wu, "Economic and Climatic Determinants of Farmer Suicide in the United States," *UC Davis*, https://escholarship.org/uc/item/65q882g8

92. Lukoye Atwoli et al., "Call for Emergency Action to Limit Global Tempera-ture Increases, Restore Biodiversity, and Protect Health," *The Lancet* 398, no. 10304 (2021): 939–941, doi:10.1016/S0140-6736(21)01915-2

93. Boya Zhang et al., "Comparison of Particulate Air Pollution from Different Emission Sources and Incident Dementia in the US," *JAMA Internal Medicine* 183, no. 10 (2023): 1080–1089, doi:10.1001/jamainternmed.2023.3300.

94. R. S. Vose et al., "Temperature Changes in the United States," in *Climate Science Special Report: Fourth National Climate Assessment*, vol. I, ed. D. J. Wuebbles et al. (U.S. Global Change Research Program, 2017), 185–206, doi: 10.7930/J0N29V45

95. Andrea Thompson, "Why Hot Overnight Temperatures Are So Dangerous," *Scientific American,* July 19, 2023, https://www.scientificamerican.com/article/why-hot-overnight-temperatures-are-so-dangerous/

96. Brian Bushard, "A Different Heatwave Warning: Online Hate—Like Violent Crime—Soars with High Temperatures, Study Suggests," *Forbes,* September 7, 2022, accessed January 3, 2025, https://www.forbes.com/sites/brianbushard/2022/09/07/a-different-heatwave-warning-online-hate-like-violent-crime-soars-with-high-temperatures-study-suggests/#:~:text=Hot%20weather%20has%20also%20been,and%208.8%25%20in%20New%20York

97. Annika Stechemesser et al., "Temperature Impacts on Hate Speech Online: Evidence from 4 Billion Geolocated Tweets from the USA," *The Lancet Planetary Health* 6, no. 9 (2022): e714-e725, doi:10.1016/S2542-5196(22)00173-5

98. "Phoenix Hit 110 Degrees on 54 Days in 2023, Setting Another Heat Record," *PBS News,* September 10, 2023, accessed November 6, 2024, https://www.pbs.org/newshour/nation/phoenix-hit-110-degrees-on-54-days-in-2023-setting-another-heat-record#:~:text=It%20was%20the%2054th%20day,more%20day%2C%E2%80%9D%20he%20said

99. Jonathan Erdman, "Heat Dome Brought Record-Breaking Temperatures," *Weather Underground,* August 26, 2023, accessed November 6, 2024, https://www.wunderground.com/article/safety/heat/news/2023-08-18-heat-dome-records-midwest-plains-south-forecast

100. "August 19–25 Historic Heatwave," *National Weather Service,* accessed November 6, 2024, https://www.weather.gov/eax/August19-25HistoricHeatwave

101. "Online Hate and Harassment: The American Experience 2023," *ADL,* June 27, 2023, accessed November 6, 2024, https://www.adl.org/resources/report/online-hate-and-harassment-american-experience-2023

102. "Moab Weather in 2023," *Extreme Weather Watch,* accessed November 6, 2024, https://www.extremeweatherwatch.com/cities/moab/year-2023

103. Jeff Masters and Bob Henson, "The Scorching Summer of 2023 Reaches 'Mind-Blowing' High Temperatures," *Yale Climate Connections,* July 17, 2023, accessed November 6, 2024, https://yaleclimateconnections.org/2023/07/the-scorching-summer-of-2023-reaches-mind-blowing-high-temperatures/#:~:text=on%20the%20Storm-,The%20scorching%20summer%20of%202023%20reaches%20'mind%2Dblowing'%20high,heatwave%20continues%20to%20roast%20Europe

104. "Tracking Toxic Algae Hot Zones," *BlueGreen Water Technologies* (blog), September 15, 2023, accessed November 6, 2024, https://bluegreenwatertech.com/post/tracking-toxic-algae-hot-zones

105. Isabelle Gain, "Hot Tub Heat: Understanding the High Ocean Temperatures Off Florida's Coast," *Thompson Earth Systems Institute, Florida Museum,* September 27, 2023, accessed November 6, 2024, https://www.floridamuseum.ufl.edu/earth-systems/blog/hot-tub-heat-understanding-the-high-ocean-temperatures-off-floridas-coast/#:~:text=In%20July%202023%2C%20a%20water,events%2C%20and%20hinder%20Florida's%20economy

106. "In Hot Water: Warming Waters Are Stressing Fish and the Fishing Indus-

try," *Climate Central*, June 26, 2019, accessed November 6, 2024, https://www.cli
matecentral.org/report/in-hot-water-warming-waters-are-stressing-fish-and-the
-fishing-industry-2019

107. "Conservation Is Patriotic in U.S. Poll Finds," *U.S. Green Chamber of Commerce*, National Survey, June 16–19, 2012, conducted by the Nature Conservancy, accessed November 6, 2024, https://usgreenchamber.com/conservation-is-patrio
tic-in-u-s-poll-finds/

Chapter 9

1. James Gustave Speth, *They Knew: The US Government's Fifty-Year Role in Causing the Climate Crisis* (MIT Press, 2021).

2. Cheryl K. Chumley, "Democrats' Obsessive Bowing to Their God of Environmentalism," *Washington Times*, August 18, 2023, https://www.washingtontimes
.com/news/2023/aug/18/democrats-obsessive-bowing-to-their-god-of-environ/

3. Richard Schiffman, "Climate Anxiety Is Widespread Among Youth—Can They Overcome It?," *National Geographic*, June 29, 2022, https://www.nationalgeo
graphic.com/environment/article/climate-anxiety-is-widespread-among-youth
-can-they-overcome-it

4. Jerome Groopman, *The Anatomy of Hope: How People Prevail in the Face of Illness* (Random House, 2004), xiv.

5. Susanne C. Moser and Carol L. Berzonsky, "Hope in the Face of Climate Change: A Bridge Without Railing," (2015), https://api.semanticscholar.org/Cor
pusID:152077284

6. Elisabeth Kübler-Ross, *On Death and Dying: What the Dying Have to Teach Doctors, Nurses, Clergy and Their Own Families* (Macmillan, 1969).

7. Steven W. Running, "The 5 Stages of Climate Grief," *Numerical Terradynamic Simulation Group Publications* 173 (2007), https://scholarworks.umt.edu/ntsg_
pubs/173

8. Groopman, *Anatomy of Hope*, 210–211.

9. Samantha K. Stanley et al., "From Anger to Action: Differential Impacts of Eco-Anxiety, Eco-Depression, and Eco-Anger on Climate Action and Wellbeing," *Journal of Climate Change and Health* 1 (2021): 100003, https://doi.org/10.1016/j.jo
clim.2021.100003

10. Anandita Sabherwal et al., "Anger Consensus Messaging Can Enhance Expectations for Collective Action and Support for Climate Mitigation," *Journal of Environmental Psychology* 76 (2021): 101640, https://doi.org/10.1016/j.jenvp.2021
.101640

11. Pam Reynolds, "The Truth About Carbon Footprints," *Conservation Law Foundation*, March 18, 2024, accessed December 9, 2024, https://www.clf.org/
blog/the-truth-about-carbon-footprints/

12. Saul Elbein, "Exxon CEO Blames Public for Failure to Fix Climate Change," *The Hill*, February 2, 2024, accessed November 6, 2024, https://thehill.com/policy
/energy-environment/4494543-exxon-ceo-blames-public-for-failure-to-fix-cli
mate-change/

13. Elliott Hyman, "Who's Really Responsible for Climate Change?" *Harvard Political Review*, January 2, 2020, accessed November 1, 2024, https://harvardpoli tics.com/climate-change-responsibility/

14. Judith A. Layzer and Sara R. Rinfret, *The Environmental Case: Translating Values into Policy* (Sage/CQ Press, 2020), 413.

15. Samantha Page, "The Most Villainous Act in History," *Cosmos*, February 12, 2019, accessed November 25, 2024, https://cosmosmagazine.com/earth/climate/the-most-villainous-act-in-the-history-of-human-civilisation-michael-e-mann-speaks-out/

16. Jonathan Chait, "Why Are Republicans the Only Climate-Science-Denying Party in the World?" *Intelligencer*, September 27, 2015, https://nymag.com/intelligencer/2015/09/whys-gop-only-science-denying-party-on-earth.html; see also David Roberts, "The GOP Is the World's Only Major Climate-Denialist Party. But Why?" *Vox*, December 2, 2015, accessed November 6, 2024, https://www.vox.com/2015/12/2/9836566/republican-climate-denial-why

17. Roxanna Bardan, "NASA Analysis Confirms 2023 as Warmest Year on Record," *NASA*, Goddard Institute for Space Studies, January 12, 2024, accessed November 6, 2024, https://www.nasa.gov/news-release/nasa-analysis-confirms-2023-as-warmest-year-on-record/

18. David G. Victor, *Global Warming Gridlock: Creating More Effective Strategies for Protecting the Planet* (Cambridge University Press, 2011), 203.

19. David G. Victor, *The Collapse of the Kyoto Protocol and the Struggle to Slow Global Warming* (Princeton University Press 2001).

20. David Roberts, "The Conceptual Breakthrough Behind the Paris Climate Treaty," *Vox*, December 15, 2015, https://www.vox.com/2015/12/15/10172238/paris-climate-treaty-conceptual-breakthrough

21. Victor, *Global Warming Gridlock*.

22. Cited in Christie Aschwanden, "A Lesson from Kyoto's Failure: Don't Let Congress Touch a Climate Deal," *FiveThirtyEight*, December 4, 2015, accessed November 6, 2024, https://fivethirtyeight.com/features/a-lesson-from-kyotos-failure-dont-let-congress-touch-a-climate-deal/

23. David G. Victor, "Why Paris Worked: A Different Approach to Climate Diplomacy," *YaleEnvironment360*, December 15, 2015, accessed November 6, 2024, https://e360.yale.edu/features/why_paris_worked_a_different_approach_to_cli mate_diplomacy; see also https://unfccc.int/process-and-meetings/the-paris-ag reement

24. The Conference of the Parties is the supreme decision-making body of the United Nations Framework Convention on Climate Change (UNFCCC), https://unfccc.int/process/bodies/supreme-bodies/conference-of-the-parties-cop

25. "Open Letter on COP Reform to All States That Are Parties to the Convention, Mr. Simon Stiell, Executive Secretary of the UNFCCC Secretariat and UN Secretary-General António Guterres," *Club of Rome*, November 15, 2024, accessed November 25, 2024, https://www.clubofrome.org/cop-reform-2024/

26. Charles F. Sabel and David G. Victor, *Fixing the Climate: Strategies for an Uncertain World* (Princeton University Press, 2022).

27. A. C. Thompson, "Timeline: The Science and Politics of Global Warming," *Frontline Politics,* accessed November 6, 2024, https://www.pbs.org/wgbh/pages/frontline/hotpolitics/etc/cron.html

28 Ben Lefebvre and Zack Colman, "Trump Would Withdraw US from Paris Climate Treaty Again, Campaign Says," *Politico,* June 28, 2024, https://www.politico.com/news/2024/06/28/trump-paris-climate-treaty-withdrawal-again-00165903

29. David Joravsky, *The Lysenko Affair* (University of Chicago Press, 1970), 306.

30. Gregg Sparkman et al., "Americans Experience a False Social Reality by Underestimating Popular Climate Policy Support by Nearly Half," *Nature Communications* 13, no. 4779 (2022), https://doi.org/10.1038/s41467-022-32412-y

31. Rachel Treisman, "A Meteorologist Got Threats for His Climate Coverage. His New Job Is About Solutions," *NPR: Morning Edition,* June 27, 2023, https://www.npr.org/2023/06/27/1184461263/iowa-meteorologist-harassment-climate-change-quits

32. Stanley et al., "From Anger to Action," 10003.

Select Bibliography

Albrecht, Glenn. *Earth Emotions: New Words for a New World*. Cornell University Press, 2019.

Arias-Maldonado, Manuel, and Zev Trachtenberg, eds. *Rethinking the Environment for the Anthropocene: Political Theory and Socionatural Relations in the New Geological Epoch*. Routledge, 2018.

Balint, Peter J., Ronald E. Stewart, Anand Desai, and Lawrence C. Walters. *Wicked Environmental Problems: Managing Uncertainty and Conflict*. Island Press, 2011.

Benson, Melinda Harm, and Robin Kundis Craig. *The End of Sustainability: Resilience and the Future of Environmental Governance in the Anthropocene*. University Press of Kansas, 2017.

Biermann, Frank, and Eva Lövbrand, eds. *Anthropocene Encounters: New Directions in Green Political Thinking*. Cambridge University Press, 2019.

Bowen, Mark. *Censoring Science: Inside the Political Attack on Dr. James Hansen and the Truth of Global Warming*. Dutton Adult, 2007.

Castree, Noel, and Bruce Braun, eds. *Social Nature: Theory, Practice and Politics*. Wiley-Blackwell, 2001.

Cunsolo, Ashlee, and Karen Landman, eds. *Mourning Nature: Hope at the Heart of Ecological Loss and Grief*. McGill-Queen's University Press, 2017.

Dobson, Andrew. *Green Political Thought*, Routledge, 2007.

Doherty, Brian, and Marius de Geus, eds. *Democracy and Green Political Thought: Sustainability, Rights and Citizenship*. Routledge, 1996.

Dowie, Mark. *Losing Ground: American Environmentalism at the Close of the Twentieth Century*. MIT Press, 1995.

Dunlap, Julie, and Susan A. Cohen. *Coming of Age at the End of Nature: A Generation Faces Living on a Changed Planet*. Trinity University Press, 2016.

Ehrenfeld, David. 2008. *Becoming Good Ancestors: How We Balance Nature, Community, and Technology*. Oxford University Press, 2008.

Goodell, Jeff. *The Water Will Come: Rising Seas, Sinking Cities, and the Remaking of the Civilized World.* Little Brown & Co., 2017.

Gottlieb, Robert. *Environmentalism Unbound: Exploring New Pathways for Change.* MIT Press, 2001.

Gottlieb, Robert. *Forcing the Spring: The Transformation of the American Environmental Movement.* Island Press, 2005.

Gunderson, Lance H., and C. S. Holling, eds. *Panarchy: Understanding Transformations in Human and Natural Systems.* Island Press, 2002.

Hansen, James. *Storms of My Grandchildren: The Truth About the Coming Climate Catastrophe.* Bloomsbury USA, 2009.

Hertsgaard, Mark. *Hot: Living Through the Next Fifty Years on Earth.* Houghton Mifflin Harcourt, 2011.

Hulme, Mike. *Exploring Climate Change Through Science and in Society: An Anthology of Mike Hulme's Essays, Interviews and Speeches.* Routledge, 2013.

Hulme, Mike. *Why We Disagree About Climate Change: Understanding Controversy, Inaction and Opportunity.* Cambridge University Press, 2009.

Kempton, Willett, James S. Boster, and Jennifer A. Hartley. *Environmental Values in American Culture.* MIT Press, 1995.

Kolbert, Elizabeth. *Field Notes from a Catastrophe.* Bloomsbury USA, 2006.

Maniates, Michael, and John M. Meyer, eds. *The Environmental Politics of Sacrifice.* MIT Press, 2010.

Meadows, Donella H., and Diana Wright, eds. *Thinking in Systems.* Chelsea Green Publishing, 2008.

Michaels, David. *The Triumph of Doubt: Dark Money and the Science of Deception.* Oxford University Press, 2020.

Moser, Susanne, and Lisa Dilling. *Creating a Climate for Change: Communicating Climate Change and Facilitating Social Change.* Cambridge University Press, 2007.

Nicholson, Simon, and Sikina Jinnah, eds. *New Earth Politics: Essays from the Anthropocene.* MIT Press, 2016.

Nicholson, Simon, and Paul Wapner. *Global Environmental Politics: From Person to Planet.* Routledge, 2014.

Orr, David, ed. *Democracy in a Hotter Time: Climate Change and Democratic Transformation.* MIT Press, 2023.

Orr, David W. *Down to the Wire: Confronting Climate Collapse.* Oxford University Press, 2009.

Princen, Thomas, Jack P. Manno, and Pamela L. Martin, eds. *Ending the Fossil Fuel Era.* MIT Press, 2015.

Ray, Sarah Jaquette. *A Field Guide to Climate Anxiety: How to Keep Your Cool on a Warming Planet.* University of California Press, 2020.

Sellers, Christopher C. *Crabgrass Crucible: Suburban Nature and the Rise of Environmentalism in Twentieth-Century America.* University of North Carolina Press, 2012.

Soule, Michael E., and Gary Lease. *Reinventing Nature?: Responses to Postmodern Deconstruction*. Island Press, 1995.

Speth, James Gustave. *The Bridge at the End of the World: Capitalism, the Environment, and Crossing from Crisis to Sustainability*. Yale University Press, 2008.

Vaughn, Jacqueline. *Green Backlash: The History and Politics of Environmental Opposition in the U.S.* Lynne Rienner, 1997.

Whitmarsh, Lorraine, Irene Lorenzoni, and Saffron O'Neill. *Engaging the Public with Climate Change: Behaviour Change and Communication*. Routledge, 2010.

Index

Abbott, Greg, 170

"Abrupt Impacts of Climate Change" (National Academy of Sciences), 50-51

acidification, 43, 46, 123, 134

acid rain: Carter and, 79; Clean Air Act and, 33, 99; denialists and, 148; environmental science and, 79, 82, 98-99; geoengineering and, 148; hope and, 181; McKibben on, 41-42; politics and, 2, 7, 12; toxicity and, 2, 7, 12, 33, 79, 82, 98-99, 148, 181; worldviews and, 33

Action Plan, 86

activists: Anthropocene and, 55; assault on environmentalism and, 63, 66, 69-74; denialists and, 130; Earth First! and, 55; ecotage and, 55; environmental science and, 92; hope and, 185, 192

advertorials, 85

aerosols, 104, 131-32, 141, 150

Agenda 21, 34-35, 68

Age of Humans. *See* Anthropocene

agriculture: Agenda 21 and, 35; Anthropocene and, 43-44, 47, 49, 51; Carson and, 1, 49, 123; DDT and, 1, 123; denialists and, 133, 146-47; geoengineering and, 133, 146-47; global warming and, 50; irreversibility and, 54; partisan divide and, 163; scientific uncertainty and, 123; topsoil and, 20, 54, 152

AIDS, 104

alarmists: assault on environmentalism and, 70; climate change and, 7; environmental science and, 7, 13, 70, 100, 153, 160; hope and, 181; partisan divide and, 153, 160; robustness and, 13; scientific uncertainty and, 120

Alaska National Interest Lands Conservation Act, 167

Albatross decline, 6

algae, 2, 47, 178

alternative facts, 174, 182

Amazon rain forest, 50, 112

American Clean Energy and Security Act, 142-43

American Clean Power, 171

American Enterprise Institute (AEI), 88, 90, 141-43

American Legislative Exchange Council, 124

American Medical Association, 122

American Petroleum Institute, 86

American Policy Center, 34

American Psychological Association, 6

Anderson, Kevin, 137

Andrews, Richard, 33

Antarctica, 50, 54, 112

Anthropocene: activists and, 55; as Age of Humans, 15, 40, 44-52; agriculture and, 43-44, 47, 49, 51; alternative names for, 109; atmosphere and, 41-43, 46, 49;

Breakthrough Institute, 104
Breitbart, 7, 97
Bricmont, Jean, 101, 105-6, 116
Brites, Alice, 128
British Petroleum (BP), 91, 145, 185
Broecker, Wallace, 50-51
Brower, David, 59
Brysse, Keynyn, 13-14
Budryk, Zack, 145
Buell, Frederick, 69
Bullard, Robert, 61
Bureau of Land Management, 67
Bureau of Reclamation, 59
Burke, Anthony, 9
Bush, George H. W.: acid rain and, 148; bi-partisanship and, 76; Clean Air Act and, 219n1; environmental extremists and, 66; global warming and, 189; United Nations Framework Convention on Climate Change and, 189
Bush, George W.: ANWR and, 167; geoengineering and, 141; junk science and, 80-81; Kyoto Protocol and, 189; Luntz and, 76, 88, 157
Buttel, Frederick, H., 103

Canada, 43, 79, 146, 177
Can Democracy Handle Climate Change? (Fiorino), 126-27
cap-and-trade system, 142-43, 149
Cape Meares State Park, 20
capitalism: denialists and, 140; dominant social paradigm (DSP) and, 38; free market and, 140 (*see also* free market); geoengineering and, 140; woke, 69
Capitalocene, 109
carbon budgets, 147-48, 150
carbon calculator, 185
carbon capture: cap-and-trade system and, 142-43, 149; geoengineering and, 131-40, 144-46, 150; partisan divide and, 172; U.S. Department of Energy and, 137
carbon dioxide: Anthropocene and, 41, 43, 50; Clean Air Act and, 141, 143; denialists and, 14, 131, 133, 138, 144; environmental science and, 84, 88; EPA and, 141, 143; geoengineering and, 14, 131, 133, 138, 144; hope and, 186-87; Keeling Curve and, 112; *Massachusetts v. EPA* and, 141; opposition to pipelines for,

137-38; record high emissions of, 186-87; Satartia incident and, 138
carbon footprint, 91, 185
Carson, Rachel: agriculture and, 1, 49, 123; attacks on, 122-23; controversy of, 122; DDT and, 1-2, 122-23, 163, 191; environmental groups and, 60; outdoorsmen reaction to, 163-64; *Silent Spring*, 1-2, 60, 62, 122-23, 163, 190, 192; songbirds and, 1, 192; truth and, 190-91
Carter, Jimmy, 63, 79, 167, 180
Castree, Noel, 113
cathedral forests, 21, 32, 66
Cato Institute, 7, 82, 88, 103, 143
CBS News, 177
Center for Biological Diversity, 141
Center for Countering Digital Hate, 92
Center for International Environmental Law, 144
Center for the Defense of Free Enterprise, 67
chemicals: Anthropocene and, 41-42, 46-47, 54; assault on environmentalism and, 73, 77; denialists and, 137, 146; environmental science and, 102, 112; estrogens, 41, 93; fertilizers, 17, 26, 146, 174; fetal development and, 41; forever, 2, 54, 77; geoengineering and, 137, 146; partisan divide and, 163; pesticides, 1, 26, 49, 75, 93, 123, 146, 163, 174; petrochemicals, 137, 146, PFAS, 54; politics and, 1-2; scientific uncertainty and, 122-23; sulfur dioxide, 131-32, 141, 143, 149; worldviews and, 25-26
Chevron, 145, 204n39
China, 169-70, 174, 189
Christianity, 7, 22, 35, 64, 70, 153-54
Cisco Systems, 173
Citizens Climate Lobby, 62
"Citizen's Guide to Climate Change, A" (American Enterprise Institute), 88
Clark, Nigel, 110
Clark Glacier, 6
Clean Air Act, 219n1; acid rain and, 33, 99; bipartisanship and, 2; carbon dioxide and 141, 143; environmental groups and, 60; Greenland's ice and, 112
clean energy: American Clean Energy and Security Act, 142-43; assault on environmentalism, 69; ecological issues, 69,

conservatives (*cont.*)

171-75, 180-86, 191, 193; denialists and, 79, 83-98, 130-31, 137, 139-44, 147-50; elites, 37, 70, 78, 89, 104, 153-54, 159-62, 179-85, 191-93; environmental groups and, 59-62, 66, 75; fossil fuels and, 70-73, 83-92, 98, 144-50; Fox News, 7, 90-91, 97, 100; hope and, 179-86, 191-93; merchants of doubt and, 16, 80-83, 88, 93, 104; moral issues and, 89; natural sciences and, 60; partisan divide and, 3, 152-78; political paradigms and, 8, 13-17; resources and, 34, 38, 63-64, 67, 82-83, 96-97; scientific uncertainty and, 116-24; siege on environmental science by, 78-107, 114; values and, 64, 71, 73; worldviews and, 29-38

"Conservatives Should Support Action to Reduce Plastic Pollution" (*Washington Examiner*), 169

constructionists: denialists and, 102-5; environmental science and, 100-9, 111, 113-15; hope and, 190; epistemic relativists and, 102; Hansson on, 101-4; introduction to social, 100-2; knowledge and, 101-2; Latour on, 104; natural sciences and, xi, 11, 101-3, 106-7, 109, 111, 114; nature of reality, 101; oversimplification of, 17; partisan divide and, 17; politicization of, 1-7; radicals and, 101; science wars and, 105-7; scientific uncertainty and, 100-9, 111, 113-15, 123; social, 11, 17, 100-7, 109, 114, 116, 190

Consumocene, 109

Contract with America, 75

conventional wisdom, xiv, 86

Conway, Erik, 31

Cook, John, 12-13, 123

Cool It (Lomborg), 91

coral reefs, 5, 47, 50, 112, 178

Cornwall Alliance, 7, 64, 70

Cortner, Hanna, 33, 36

Cotgrove, Stephen, 31, 37

COVID-19, 49, 91, 99

Crawford Lake, 43

Cronon, William, 103

Crutzen, Paul, 41, 131-33

Cruz, Ted, 35

cryosphere, 42, 46, 113

Cuccinelli, Ken, 94

Cunsolo, Ashlee, 5

Cuyahoga River, 2, 62

Dannenberg, Astrid, 135

dark money, 88

Davis, Adam, 25-28

DDT: agriculture and, 1, 123; Carson and, 1-2, 122-23, 163, 191; dead rivers and, 163; ecological issues of, 1-2, 60, 122-23, 163, 191; effects on birds, 1-2, 60; Environmental Defense Fund (EDF), 60; US ban on, 122-23

Death Valley, 178

decision-making: Anthropocene and, 54, 56; denialists and, 136; geoengineering and, 136; policymakers and, xiii, 12, 127-28; politics and, 12-14; scientific uncertainty and, xiii-xiv, 125-28; worldviews and, 33

deconstructionists: Anthropocene and, 107-10; denialists and, 99, 102-6, 115, 120-21, 125; Earth System and, 111-15; environmental science and, 99-115; natural sciences and, 101, 106-7, 109, 111; scientific uncertainty and, 17, 116-29

Deepwater Horizon, 147

deflection, 91-92

deforestation, 7, 98, 146

Delanty, Gerard, 127

Demeritt, David, 106

democracy: assault on environmentalism and, 77; Bill of Rights, 56-57; environmental science and, 175-78; partisan divide and, 175-78; scientific uncertainty and, 125-29; threats to, 1

democratic norms: Anthropocene and, 16, 52, 54, 56-57; assault on environmentalism and, 71, 76; denialists and, 138, 141; environmental science and, 89; geoengineering and, 138, 141; hope and, 180-81; partisan divide and, 178; politics and, 10, 16; scientific uncertainty and, 17, 125

Democrats: assault on environmentalism and, 69; Carter, 63, 79, 167, 180; Clinton, 75-76, 187, 189; denialists and, 138, 148; environmental science and, 89; fossil fuels and, 138, 172, 180-81, 185; geoengineering and, 138, 148; hope and, 180-81, 185, 188; Obama, 117, 142-44, 180, 189; partisan divide and, 3, 153, 160-61, 164, 167-69, 172

ronmental science and, 84, 113; famine and, 133; GMO, 190; hope and, 190; partisan divide and, 173-74; politics and, 1; reducing waste of, 173-74

Forbes, 177

forests: Anthropocene and, 41-44, 47, 50, 53, 55; assault on environmentalism and, 59, 64-68; birch, 26; cathedral, 21, 32, 66; complex adaptive systems (CAS) and, 26-27; deforestation and, 7, 98, 146; denialists and, 131, 146; Douglas Fir, 26, 65; Earth First! and, 55; environmental science and, 82, 98, 102, 112-14; feller bunchers and, 66; fertilizers and, 26; fungi and, 25-26, 32; geoengineering and, 131, 146; grapple skidders and, 66; health and, 25, 146; logging of, 20-21, 53, 66, 68; mother trees, 20; old-growth, 20-21, 31, 53, 55, 65-68; Pacific Northwest, 20, 31, 53, 55, 65-67, 178; partisan divide and, 163, 167-68; pesticides and, 26; rain, 50, 112; root disease and, 26; timber industry and, 20-21, 31-32, 47, 53, 55, 64-68; tropical, 43; U.S. Forest Service, 31-32, 55, 59, 66-67; wildfires and, 5, 48, 74, 90, 154, 176-77; Wise Use Movement and, 67-68; worldviews and, 20-21, 25-27, 31-32

forever chemicals, 2, 54, 77

fossil fuels: addiction to, 186; Agenda 21 and, 34-35, 68; American Petroleum Institute, 86; Anthropocene and, 41, 49; assault on environmentalism and, 70-73; Carter and, 180; climate change and, 13, 17-18, 83-91, 98, 117, 130, 133, 138-42, 145-47, 150, 166, 172-75, 180-81, 185-86, 191; conservatives and, 70-73, 83-92, 98; *Deepwater Horizon* accident and, 147; Democrats and, 138, 172, 180-81, 185; denialists and, 83-90, 130, 133-34, 138-50; environmental science and, 83-92, 98; geoengineering and, 130, 133-34, 138-50; global energy market, 147; hope and, 180-81, 185-88, 191; industry fabrications and, 83-86; moral issues and, 89; Obama and, 180; partisan divide and, 166, 170-75; plastics and, 41 (*see also* plastics); politics and, 4, 13, 17-18; Republicans and, 70, 138-40, 172, 174, 180-81, 185-86; role of industry in,

144-46; scientific uncertainty and, 117-18; Speth on, 180; worldviews of, 34, 146-50

Fossil Future (Epstein), 89

Fox, Nicholas, 103

Fox News, 7, 90-91, 97, 100

Franta, Benjamin, 87, 145

free market: assault on environmentalism and, 64; Cornwall Alliance and, 7; denialists and, 130, 140; environmental science and, 83; geoengineering and, 130, 140; paradigms and, 22; partisan divide and, 158, 162; worldviews and, 22, 33, 35, 38

Fremaux, Anne, 48-49, 52

freshwater, 2, 95, 178

Fuller, Steve, 103-4

fungi, 25-26, 32

Gallup polls, 65, 72, 153, 167

Gateway Belief Model (GBM), 158

Gayon, J., 101

Geiger, Nathaniel, 161

geoengineering: acid rain and, 148; aerosols and, 104, 131-32, 141, 150; agriculture and, 133, 146-47; Anthropocene and, 45, 50-52; atmosphere and, 131-33, 137, 139, 146, 150; biodiversity and, 135, 146; bipartisanship and, 148; capitalism and, 140; carbon capture and, 131-40, 144-46, 150; carbon dioxide and, 14, 131, 133, 138, 144; chemicals and, 137, 146; coal and, 130, 140, 144-46, 148-49; conservation and, 149; conservatives and, 17, 130-31, 139-44, 147-50, 180-81; Crutzen and, 131-33; decision-making and, 136; democratic norms and, 138, 141; Democrats and, 138, 148; denialists and, 17-18, 45, 50-52, 130-50, 180-81; dominant social paradigm (DSP) and, 140; early warning systems for, 136; Earth System and, 146; ecological issues and, 137, 146, 149-50; economic issues and, 135-37, 140, 147-50; ecosystems and, 146-47; elites and, 136, 145-46; environmentalists and, 131, 148; extinction and, 133; fear and, 132, 142; floods and, 133; forests and, 131, 146; fossil fuels and, 130, 133-34, 138-50; free market and, 130, 140; geoengineering and, 180-81; global

scientific uncertainty and, 117; sulfur dioxide, 131-32, 141, 143, 149

Greenland, 13-14, 50, 54, 110, 112

Green New Deal, 90

grief: ecological, 5-7; hope and, 179, 183-85; mass consumerism and, 5; partisan divide and, 164, 175; solastalgia, 5, 164, 183

Groopman, Jerome, 4, 184

Gross, Paul, 105

Grumet, Jason, 171

Guardian, The, 96, 145-46

Gulf of Mexico, 54, 147

Gulf Stream, 51

Gunaratnam, Yasmin, 110

Gustafson, Abel, 163

Guterres, António, 120, 130, 148

habitat loss, 2, 163-64

Hamilton, Clive, 9, 42, 56

Hansen, James, 84, 140, 186

Hansson, Sven Ove, 101-4

Harrington, Cameron, 10

Harvard's Solar Geoengineering Research Program, 145

hate speech, 178

Haugen, Francis, 92

Hayes, Tyrone, 93

health: Anthropocene and, 50, 56; assault on environmentalism and, 60-63, 71-77; denialists and, 132, 138, 140, 143, 146-47, 150; environmental science and, 80, 82, 88, 95; forests and, 25, 146; geoengineering and, 132, 138, 140, 143, 146-47, 150; hope and, 185, 191; oil spills and, 147; partisan divide and, 172-74, 177; politics and, 2, 5-6, 16, 18; scientific uncertainty and, 117-18; worldviews and, 25

Heartland Institute, 7, 82, 88-89, 97, 120, 142

heat domes, 177

Heated Debate, The: Greenhouse Predictions Versus Climate Reality (Balling), 103-4

heat waves, 2, 90, 154, 177-78

Hedlund-de Witt, Nicholas, 10

Heilbroner, Robert, 126

Hochachka, Gail, 28

Hoffman, Andrew, 155

Holcomb, Eric, 171

Holocene, 41, 44-45, 48

hope: acid rain and, 181; activists and, 185, 192; alarmists and, 181; anger and, 185-86; anxiety and, 19; authentic, 4, 6, 18, 179, 182-85, 192-93; biodiversity and, 179-83, 190; bipartisanship and, 19, 179-80, 184, 192-94; carbon dioxide and, 186-87; climate change and, 179-93; coal and, 180; conservatives and, 179-86, 191-93; constructionists and, 190; democratic norms and, 180-81; Democrats and, 180-81, 185, 188; denialists and, 180-86, 189, 191-93; dominant social paradigm (DSP) and, 179-82, 185; ecological issues and, 180-83, 192-95; economic issues and, 180-82, 186, 188; elites and, 179-85, 191-93; environmentalists and, 181, 192; extremists and, 181; fear and, 184; food and, 190; fossil fuels and, 180-81, 185-88, 191; global warming and, 180-92; greenhouse gases (GHGs) and, 186-89; grief and, 179, 183-85; health and, 185, 191; ideologies and, 180; Intergovernmental Panel on Climate Change (IPCC) and, 183; media and, 184, 186, 193; moral issues and, 192; negotiation and, 187-89; new environmental paradigm (NEP) and, 192; optimism and, 4; ozone and, 181; paradigms and, 179-80, 192; Paris Agreement and, 187-89; partisan divide and, 182, 185; plastic and, 186; politics and, 4-8; public opinion and, 193; radicals and, 187, 192; religion and, 190; renewable energy and, 187-88; Republicans and, 180-91; resources and, 181; scientific consensus and, 181, 190; security and, 182, 188; *Silent Spring* and, 190, 192; social change and, 185, 193; solar energy and, 180; stability and, 182, 192, 194; sustainability and, 192; truth and, 4, 6, 18, 179, 181-85, 189-93; water and, 193; worldviews and, 180-82, 185, 192-93

horticulture, x

Houghton Mifflin, 123

House Committee on Natural Resources, 96-97

House Committee on Science and Technology, 142

House Science Committee, 162

environmental paradigm (NEP)); new ecological paradigm statements and, 195; partisan divide and, 151-53, 179-82; politics and, 8-10, 15, 18, 28-35; resources and, 181; scientific uncertainty and, 70-77, 120-21; shifts in, 32-39; values and, 8, 22, 29; worldviews and, 21-25, 28-38

Paris Agreement: geoengineering and, 133, 137, 144; hope and, 187-89; Kyoto Protocol and, 187-89; Obama and, 189; partisan divide and, 171; pledge and review framework of, 188; Trump and, 143-144, 189

partisan divide: agriculture and, 163; alarmists and, 153, 160; anxiety and, 177; Arctic National Wildlife Refuge (ANWR) and, 167-68; assault on environmentalism and, 76; atmosphere and, 154-55; barriers of, 153-55; biodiversity and, 153, 159, 163-64, 168-69, 172, 174-75, 177; bipartisanship and, 18, 152-53, 160, 167-70, 174-75, 178; bridging, 151-78; carbon capture and, 172; chemicals and, 163; climate change and, 3, 153-78; coal and, 169, 172; cognitive predispositions and, 155; conservation and, 151, 163-69, 175; conservatives and, 3, 152-78; constructionists and, 17; democracy and, 175-78; democratic norms and, 178; Democrats and, 3, 153, 160-61, 164, 167-69, 172; denialists and, 7-8, 16-18, 152-64, 174-75; dominant social paradigm (DSP) and, 156; ecological issues and, 151-52, 156, 163-64; economic issues and, 152, 169-77; elites and, 153-55, 159, 161-62; emotion and, 18, 163-64, 177; entering conversations through back door, 165-75; environmentalists and, 3, 7, 152-55, 160, 165-66, 168, 175; environmental science and, 89; equity and, 173; exemptionalism and, 152, 160; expertise and, 151, 197n2; extinction and, 159, 164-65, 174; extremists and, 153, 160; fear and, 160-61, 166, 175; floods and, 154, 176-77; food and, 173-74; forests and, 163, 167-68; fossil fuels and, 166, 170-75; free market and, 158, 162; gerrymandering and, 76; global warming and, 154-72, 175-77; governance and, 173;

greenhouse gases (GHGs) and, 169, 173; grief and, 164, 175; health and, 172-74, 177; hope and, 182, 185; Independents and, 167-69; labels and, 167, 170; liberals and, 3, 152-78; manufactured, 154-55, 159; natural sciences and, 11; negotiation and, 165; opportunities for, 155-75; paradigms and, 151-53, 179-82; Paris Agreement and, 171; pesticides and, 163, 174; plastic and, 153, 169, 172-75; pluralistic ignorance, 161; pollution and, 153, 160, 169, 172-75; public opinion and, 156, 159, 164, 166, 169; radicals and, 155; Reagan and, 167; religion and, 153, 162, 178; Republicans and, 3, 153, 159-64, 167-74; resources and, 152, 173-74, 178; scientific consensus and, 156-61, 174; security and, 162, 172, 174, 176; self-silencing and, 161, 166; sharing stories and, 161-64; *Silent Spring* and, 163; social change and, 156, 175; solar energy and, 170-71, 173; stereotypes and, 152; sustainability and, 151, 172-75; toxicity and, 153, 178; Trump and, 153, 160, 164, 167, 171; values and, 162, 166-67, 169, 171; water and, 160, 168-69, 174, 178; worldviews and, 151-60, 174-75

Pender, Anne, 28

Persson, Johannes, x-xi

pesticides: Anthropocene and, 49; assault on environmentalism and, 75; DDT, 1-2, 60, 122-23, 163, 191; denialists and, 146; environmental science and, 93; forests and, 26; geoengineering and, 146; partisan divide and, 163, 174; politics and, 1-2; scientific uncertainty and, 123

petrochemicals, 137, 146

petroleum, 74, 86, 91, 130, 145-46, 174

Pew Research Center, 3, 6, 62, 69, 72, 170, 174

PFAS, 54

physical reality, ix

Pinchot, Gifford, 31, 67

Pirages, Dennis, 39

Planetary Era: Anthropocene and, 40, 45-52, 57; Earth System and, 40, 46, 52, 57; unknown unknowns of, 51-52

Planet Gore blog, 90

plastic: Anthropocene and, 41; assault on environmentalism and, 73-77; bag bans

plastic *(cont.)*
  and, 76; Beyond Plastics, 169; denialists
  and, 146; forever chemicals and, 2; geoengineering and, 146; hope and, 186;
  mass consumerism and, 5; microplastics, 172; partisan divide and, 153, 169,
  172-75; politics and, 2, 6, 18; pollution
  and, 6, 18, 153, 169, 172, 174-75; pollution by, 6, 18, 153, 169, 172, 174-75
polarization, 1, 76, 89, 153, 159, 170
policymakers: credible science and, xiii-
  xiv; decision-making and, xiii, 12, 127-
  28; denialists and, 132; environmental
  science and, 80, 82, 87-88; geoengineering and, 132; politics and, 12, 17; scientific uncertainty and, xiii-xiv, 116-18,
  121-24, 127-29
policy reversibility, xiv, 10, 56, 125
political consensus, 17, 106
Politico, 171
politics: acid rain and, 2, 7, 12; anxiety and,
  5-6; biodiversity and, 1, 6-7, 9, 12-13,
  18-19; chemicals and, 1-2; climate
  change and, 2-19; coal and, 10; conservation and, 4, 18; conservatives, 8, 13-17
  (see also conservatives); constructionists
  and, 1-7; controversy and, 121-23;
  decision-making and, 12-14; democratic
  norms and, 10, 16; denialists and, 130-
  50; Earth System and, 10, 15; ecological
  issues and, 1-2, 5-10, 15-18; economic
  issues and, 8, 12, 15; ecosystems and, 5,
  10, 12; emotion and, 4-8; environmentalists and, 2-3, 7-9, 16; environmental
  science and, 11-14, 82-83; exemptionalism and, 8, 15; extinction and, 2, 7; extremists and, 7; false equivalencies and,
  3, 18, 124, 180, 184, 194; food and, 1;
  fossil fuels and, 4, 13, 17-18; geoengineering and, 17-18, 130-50; gerrymandering and, 76; global warming and, 1-9,
  12-14, 17; governance and, 16; honesty
  in, 4, 179, 191, 193; ideologies and, xii, 3,
  8; Intergovernmental Panel on Climate
  Change (IPCC) and, 12-14; irreversibility and, 10; left-wing cause and, 7-8; legitimacy and, 7, 11, 17; media and, 3, 6-7,
  15-16; new story needed for, 1; Oreskes
  and, 10-12; ozone and, 2, 7, 14; paradigms and, 8-10, 15, 18, 28-35; partisan

divide and, 151-78; pesticides and, 1-2;
  Pew Research Center and, 3, 6; plastic
  and, 2, 6, 18; polarized, 1, 76, 89, 153,
  159, 170; policymakers and, 12, 17; pollution and, 1-2, 5-6, 18; radicals and, 55,
  58; scientific consensus and, 12-14, 17,
  157-58; scientific uncertainty and, 125-
  29; social sciences and, 9; stability and,
  1, 15; sustainability and, 5, 18; toxicity
  and, 2; values and, 8, 13; water and, 2, 4;
  worldviews and, 8-9, 15, 18, 28-32
pollution: acid rain, 2, 7, 12, 33, 79, 82, 98-
  99, 148, 181; aerosols and, 104, 131-32,
  141, 150; air, 2, 5, 74, 112, 118, 132, 141,
  146, 169, 172, 174; Anthropocene and, 41,
  54; assault on environmentalism and,
  60-62, 69, 74-75; collision course of, xi-
  xii; coral reefs, 5, 47, 50, 112, 178;
  Cuyahoga River fire, 2, 62; denialists
  and, 132, 141, 147; environmental science and, 82; fertilizers, 17, 26, 146, 174;
  fine particulates, 117-18; forever chemicals, 2, 54, 77; geoengineering and, 132,
  141, 147; GHGs and, 186-89 *(see also*
  greenhouse gases (GHGs)); industrialization and, 43, 72; Kyoto Protocol and,
  85-86; lifestyle choices and, 34; litter, 2,
  41, 62; mining and, 53, 67, 146; partisan
  divide and, 153, 160, 169, 172-75; pesticides, 1, 26, 49, 75, 93, 123, 146, 163, 174;
  plastic, 6, 18, 153, 169, 172, 174-75; politics and, 1-2, 5-6, 18; scientific uncertainty and, 117-18; water, 2, 32, 60, 62,
  82, 160; worldviews and, 24, 32-34
Pollution Prevention Act, 33
"Pollution Prevention Pays" (3M initiative),
  173
Pomerance, Rafe, 187
PragerU, 89
Project Drawdown, 174
property rights, 34, 138, 172
protest, 2, 35, 95, 147
PTSD, 191
public opinion: assault on environmentalism and, 16, 63-64, 69, 76; climate
  change and, 160-61; environmental science and, 100; hope and, 193; partisan
  divide and, 156, 159, 164, 166, 169; worldviews and, 25, 28, 35

INDEX

Rabe, Barry, 170-71

Radder, Hans, 104

radicals: Anthropocene and, 55, 58; assault on environmentalism and, 64-65, 68, 70; constructionists and, 101; hope and, 187, 192; left-wing, 7, 58, 64-65, 78, 90; partisan divide and, 155; politics and, 55, 58; siege on environmental science and, 78, 90, 97, 101

rationality, 13

Reagan, Ronald, 33; assault on environmentalism and, 63-66, 75-76; drilling on public lands, 167; partisan divide and, 167; siege on environmental science and, 79

"Recalculating the Climate Math" (McKibben), 148

recycling, 173, 186

"Regards State Higher Ed Institution Commitment to Certain Beliefs" (HB 394), 125

Regional Greenhouse Gas Initiative (RGGI), 149

"Relationship Between Systems Thinking and the New Ecological Paradigm, The" (Davis and Stroink), 25

religion: assault on environmentalism and, 16, 70; changing Almighty's creation, 153; Christianity, 7, 22, 35, 64, 70, 153-54; environmental science and, 153; evangelicals, 7, 16, 58, 64, 70, 153-54; hope and, 190; lived experience and, 154-55; partisan divide and, 153, 162, 178; scientific uncertainty and, 124; worldviews and, 8-9

renewable energy: environmental science and, 85, 92; hope and, 187-88; partisan divide and, 170-71, 174; sustainability and, 23, 85, 92, 170-71, 174, 187-88; worldviews and, 23

Republican National Committee, 35

Republicans: assault on environmentalism and, 63-64, 69-70, 72, 75-76; Bush administrations, 148 (see also Bush, George H. W.; Bush, George W.); denialists and, 138-40, 148; Dirty Water Act and, 75; elites, 70, 89, 159-62, 181-85; environmental science and, 81, 87, 89, 96, 100; fossil fuels and, 70, 138-40, 172, 174, 180-81, 185-86; geoengineering and,

138-40, 148; hope and, 180-91; merchants of doubt and, 16, 80-83, 88, 93, 104; Nixon, 63, 75; partisan divide and, 3, 153, 159-64, 167-74; Reagan, 33, 63-66, 75-76, 79, 167; "The Specter of Environmentalism", 63; Trump and, 69 (see also Trump, Donald); worldviews and, 35

republicEn, 161-62, 166, 181

Resisting the Green Dragon (Cornwall Alliance), 70

resources: Anthropocene and, 41, 44, 52, 55-56; assault on environmentalism and, 60, 63-67, 77; biodiversity, 1 (see also biodiversity); conservatives and, 34, 38, 63-64, 67, 82-83, 96-97; deforestation, 7, 98, 146; environmental science and, 82-83, 96-97; fossil fuels, 41 (see also fossil fuels); hope and, 181; limited, 22, 38, 152, 195; paradigm statements and, 181; partisan divide and, 152, 173-74, 178; pollution and, 1-2 (see also pollution); warning to humanity on, xi-xii; worldviews and, 22, 26-38

reversibility, xiv, 10, 54, 56, 125

Reynolds, Kim, 170-71

Richocene, 109

Riders from Hell (*Washington Post*), 75-76

Rio Declaration on Environment and Development, 34-35

"Rise of Extinction Deniers" (*Scientific American*), 96

rivers: Colorado River, 59; Cuyahoga River fire, 2, 62; DDT and, 163; dead, 163; ecosystems and, 2, 20, 59, 62, 163, 178; global warming and, 178; pollution of, 2, 62; worldviews and, 20

Roberts, David, 139

robustness: alarmists and, 13; Anthropocene and, 43; biodiversity and, 174; environmental science and, 104; global warming and, 188; scientific uncertainty and, 116, 123; worldviews and, 25

Rolling Stone, 147-48

Roosevelt, Theodore, 164, 167

Rosenbaum, Walter, xiii, 118, 121, 128

Ruckelshaus, William, 33

Ruddiman, William, 43

Running, Steven, 183

Russia, 147

worldviews (*cont.*)

and, 8-9, 15, 18, 28-32; pollution and, 24, 32-34; public opinion and, 25, 28, 35; religion and, 8-9; renewable energy and, 23, 170-71, 174; Republicans and, 35; resources and, 22, 26-38; rivers and, 20; robustness and, 25; scientific consensus and, 25, 156-61; scientific uncertainty and, 123; social change and, 31-32, 37; social paradigms, x, 8, 15, 21-22, 36, 79, 120-21, 151, 180, 192, 195; social sciences and, 38; sustainability and, 21, 28, 32, 34-35; values and, 22, 25, 28-29, 31, 34-35; water and, 26, 32

Wynne, Brian, 103

X, 92

Xiao, C., 24

Yale Program on Climate Change Communication, 72, 157

Yearley, Steven, 103

Yellowstone National Park, 168

Young, Don, 70

YouTube, 89, 92

Zitzelsberger, Sonja, 135